Introduction to Continuum Biomechanics

Introduction to Continuum Biomechanics
Kyriacos A. Athanasiou and Roman M. Natoli

ISBN: 978-3-031-00498-8 paperback

ISBN: 978-3-031-01626-4 ebook

DOI: 10.1007/978-3-031-01626-4

A Publication in the Springer series
SYNTHESIS LECTURES ON BIOMEDICAL ENGINEERING #19

Lecture #19

Series Editor: John D. Enderle, University of Connecticut

Series ISSN

ISSN 1930-0328 print

ISSN 1930-0336 electronic

Introduction to Continuum Biomechanics

Kyriacos A. Athanasiou and Roman M. Natoli
Rice University

SYNTHESIS LECTURES ON BIOMEDICAL ENGINEERING #19

ABSTRACT

This book is concerned with the study of continuum mechanics applied to biological systems, i.e., continuum biomechanics. This vast and exciting subject allows description of when a bone may fracture due to excessive loading, how blood behaves as both a solid and fluid, down to how cells respond to mechanical forces that lead to changes in their behavior, a process known as mechanotransduction. We have written for senior undergraduate students and first year graduate students in mechanical or biomedical engineering, but individuals working at biotechnology companies that deal in biomaterials or biomechanics should also find the information presented relevant and easily accessible.

KEYWORDS

Continuum mechanics, biomechanics, elasticity, Newtonian fluids, blood flow, Casson equation, linear viscoelasticity, poroelasticity, thermoelasticity, biphasic theory

Dedication

To Kiley, Aristos, and Thasos for being the center and providing the apogee of my life. To all of my graduate students for sharing the passion for the unrelenting pursuit of excellence. -KA2

To God the father, for the gift of life, and my earthly father, the late Joseph D. Natoli, for instilling in me a love of science and mathematics. Hopefully, these two are together now. -RMN

Foreword

This book has evolved from notes that have been developed for a first-year graduate course in biomechanics at Rice University. The material is presented in the same order that is usually taught, and is easily covered in a standard, semester-long, three credit hour course. In this book, we first develop the necessary mathematical, kinematic, and stress analysis concepts essential to continuum theory (Chapters 1–3). We assume no knowledge of tensor algebra or calculus, as we present everything needed to follow this book. However, a basic understanding of elementary mechanics, vector algebra and calculus, multivariable calculus, and differential equations is helpful.

In Chapter 4, we begin with elasticity, as it is the archetypical constitutive theory. We first cover general isotropic elasticity, devoting some time to hyperelastic descriptions of finite deformation, as applied to *tendons* or *ligaments*. We then address linear elasticity, detailing the experimental observations supporting this model and solving some of the classical problems, using *bone* as an example. Chapters 5 and 6 deal exclusively with fluids. While non-Newtonian fluids are briefly discussed, the main focus of Chapter 5 is Newtonian fluids, which are described by a linear constitutive equation. We depart from linearity in Chapter 6, where the Casson equation describing *blood flow* is studied in its different regimes, and the equations of motion are solved for flow within a large cylindrical vessel. In Chapter 7, we return to linear theory, discussing linear viscoelasticity. Both integral and differential (i.e., spring-dashpot) descriptions are presented. Furthermore, the correspondence principle and methods for analyzing dynamic loading of viscoelastic materials are discussed. Here, our biological examples are *tendons/ligaments* and a *single cell*. In Chapter 8, we briefly look at poroelasticity and thermoelasticity, which are examples of coupled theories. Finally, Chapter 9 deals with the subject of mixture theory. We devote the majority of the chapter to biphasic theory, as applied to *articular cartilage* and *fibrocartilage* biomechanics, solving both creep and stress relaxation problems in confined compression and deriving the governing equation for unconfined compression.

We hope you find the material comprehensive enough, but not overly bogged down with mathematics. Within each chapter, we give examples, which we call demonstrations, of important concepts. In addition, each chapter is followed by several problems that reinforce or expand upon the material presented. For the more interested reader, references to classical papers and more advanced textbooks on both mechanical and biological aspects of the subject matter are provided.

Of course, we are excited by the subject of continuum biomechanics and hope that you share our enthusiasm by the time you finish this book!

A note to instructors: sections and chapters preceded by an asterisk in the table of contents can be omitted from a one semester course without loss of continuity. Also, a solutions manual is available upon request.

Contents

Introduction

CONTINUUM THEORY

We know that matter is composed of discrete units called molecules, which consist of atoms and subatomic particles. Thus, matter is not continuous. In principle, we could describe the macroscopic behavior of materials based on detailed knowledge of the microscopic behavior using statistical mechanics, though even the simplest of problems can be mathematically intractable. However, and fortunately, many aspects of daily life can be described or predicted with theories that pay no attention to the molecular structure of materials (e.g., bone deformation, blood flow). Continuum theory describes relationships between phenomena by paying no attention to the molecular structure of materials. It regards matter as indefinitely divisible, an assumption that is justified only so far as it can adequately capture experimental observations.

CONTINUUM MECHANICS

Mechanics is the branch of physics relating applied forces to motion. Continuum mechanics studies the response of materials, both solids and fluids (including liquids and gases), at the macroscopic level when subject to different loading conditions. While rigid body motion is included in this theory as a special reduction, deformation of materials under applied loads is its most common application. The subject of continuum mechanics can be subdivided into two primary parts: general principles (i.e., governing equations) and constitutive relationships. *General principles* are common to all materials (e.g., conservation of mass and energy, principle of linear momentum and angular momentum, principle of entropy, etc.) These governing equations can take two different, yet equivalent, forms. One is an integral formulation over a finite volume. The second is field equations for a particle or differential material volume at every point. In this text, we consider only the latter.

Constitutive relationships are equations defining idealized materials. Some examples are an elastic body (deformation → 0 as load → 0), linearly elastic solid (stress linearly related to strain), and linearly viscous fluid (stress linearly related to strain rate). In general, constitutive equations vary from material to material and, for a given material, may vary depending on loading conditions.

For example, articular cartilage exhibits behavior reminiscent of an incompressible hyperelastic solid under quick-loading conditions, such as may occur during impact, but acts as a linear biphasic material under quasi-static loading conditions common to confined, unconfined, and creep indentation testing. We will investigate aspects such as these, and many other constitutive equations, over the course of this text.

The elegance of continuum mechanics is its mathematical framework that yields concise treatment of the governing equations and constitutive relationships. In continuum mechanics two aspects of investigation can be considered. One is the formulation of constitutive equations appropriate for describing behaviors of particular classes of materials under certain loading conditions. The second is solving the governing equations, in combination with the constitutive equations, under circumstances of interest to scientists and engineers. In this text, we seek to develop the necessary tools to work on both aspects at an introductory level. Explicit rules for constitutive equation formulation (e.g., coordinate invariance, determinism, and material objectivity, etc.) [1] and advanced methods for solving particular problems are left for an intermediate text focused on more specialized applications.

CONTINUUM BIOMECHANICS

Biomechanics is the application of engineering mechanics to biological systems. It is important to keep in mind that your study of continuum biomechanics is rich in the sense that the majority of equations and problem solving techniques you will learn are applicable to a large range of engineering materials and not just to human locomotion or to tissues and cells within the body. Biomechanics is suitable for research and development, as well as design and analysis, of problems having biological or medical interest. In this text, we present continuum mechanics applied to biological systems, or continuum biomechanics. Continuum biomechanics is part of a larger field of mechanobiology, which is the study of the response of biological systems to mechanical forces. As a framework for mechanobiology, one can think of two distinct, yet linked, responses. The first response is the mechanical response of the system, which is determined by governing equations, constitutive relationships, and failure theories, to name a few. The second response is a biological one. This response is determined by the history of *mechanotransduction*, the process by which mechanical signals initiate gene transcription, leading to protein production, and ultimately affecting a tissue's structure–function relationship.

As the term biomechanics implies, there is an inherent treatment of biology in addition to the description of mechanics principles. This book concentrates on the mechanics of empirically observed relationships between stress and strain or strain rate, with examples taken from biological systems. We do not concentrate on the anatomy, physiology, or structure–function relationships of the tissues of interest, as we expect this knowledge to have been obtained through other means.

It has been said that a bioengineer is one who takes a biological problem and reduces it, under appropriate assumptions, to a problem in one of the engineering disciplines. No where may this be truer than in the role of a biomechanician. With this said, let us get started!

CHAPTER 1

Tensor Calculus

1.1 INDICIAL NOTATION

Continuum mechanics is a mathematically elegant theory. As such, we must be able to concisely express the equations governing and describing the behavior of materials. We will see shortly that indicial notation simplifies the bookkeeping process of the many mathematical expressions needed in continuum mechanics by allowing us to write them in a compact form.

1.1.1 Summation (Dummy Indices)

Consider the sum

$$S = a_1 x_1 + a_2 x_2 + a_3 x_3 + \ldots + a_n x_n \tag{1.1}$$

or

$$S = \sum_{i=1}^{n} a_i x_i \tag{1.2}$$

or

$$S = \sum_{j=1}^{n} a_j x_j = \sum_{k=1}^{n} a_k x_k \tag{1.3}$$

The indices i, j, k in Eqs. (1.2) and (1.3) are called dummy indices. Thus, if an index is repeated once within a simple term, it is a *dummy index* indicating a summation. By convention, each dummy index has values 1, 2, 3. For example,

$$S = \sum_{i=1}^{3} a_i x_i \qquad \text{(Sum of 3 terms)}$$

and (1.4)

$$S = \sum_{i=1}^{3} \sum_{j=1}^{3} a_{ij} x_i x_j \qquad \text{(Sum of } 3^2 \text{ terms)}$$

To alleviate writing the summation symbol continuously, we will employ Einstein's summation convention. With this convention, it is understood that the summation occurs over repeated indices without having to write the summation symbol. Thus, Eqs. (1.4) become

$$S = a_i x_i$$

and (1.5)

$$S = a_{ij} x_i x_j$$

According to this convention, expressions such as $S = a_i b_i c_i$ (where i is repeated more than two times) are meaningless.

1.1.2 Multiple Equations (Free Indices)

Consider

$$\begin{aligned} y_1 &= a_{11}x_1 + a_{12}x_2 + a_{13}x_3 \\ y_2 &= a_{21}x_1 + a_{22}x_2 + a_{23}x_3 \\ y_3 &= a_{31}x_1 + a_{32}x_2 + a_{33}x_3 \end{aligned} \tag{1.6}$$

Using dummy indices, we can reduce Eqs. (1.6) to

$$\begin{aligned} y_1 &= a_{1n}x_n \\ y_2 &= a_{2n}x_n \\ y_3 &= a_{3n}x_n \end{aligned} \tag{1.7}$$

Notice Eqs. (1.7) are of the same form, differing only by the index 1, 2, 3. Thus, by introducing the concept of a free index, Eqs. (1.7) can be written as

$$y_i = a_{in}x_n \tag{1.8}$$

where i is the free index and n is a dummy index (as described in Section 1.1.1). *Free indices* appear only once in each term of an equation, indicating multiple equations. They typically take on the values 1, 2, 3.

For example, $y_i = a_{im} x_m$ represents three equations, each with three terms on the right-hand side. The free index appearing in every term or equation must be the same. If there are two free indices in an equation, then the expression represents nine equations. For example,

$$T_{ij} = A_{im} A_{jm} \tag{1.9}$$

represents nine equations, each with three terms on the right-hand side. Taking the free indices $i = 1, j = 2$,

$$T_{12} = A_{11}A_{21} + A_{12}A_{22} + A_{13}A_{23} \tag{1.10}$$

There are eight other equations for the other combinations of i and j.

1.1.3 Kronecker Delta

The Kronecker delta is defined as

$$\delta_{ij} = \begin{cases} 1 & i = j \\ 0 & i \neq j \end{cases} \tag{1.11}$$

i.e., $\delta_{11} = \delta_{22} = \delta_{33} = 1$ and $\delta_{12} = \delta_{13} = \delta_{21} = \ldots = 0$. The matrix

$$\begin{bmatrix} \delta_{11} & \delta_{12} & \delta_{13} \\ \delta_{21} & \delta_{22} & \delta_{23} \\ \delta_{31} & \delta_{32} & \delta_{33} \end{bmatrix} = \begin{bmatrix} 1 & 0 & 0 \\ 0 & 1 & 0 \\ 0 & 0 & 1 \end{bmatrix} \tag{1.12}$$

is the identity matrix. Now let us consider some examples:

1.
$$\delta_{ii} = \delta_{11} + \delta_{22} + \delta_{33} = 3 \tag{1.13}$$

2.
$$\begin{aligned} \delta_{1m} a_m &= \delta_{11} a_1 + \delta_{12} a_2 + \delta_{13} a_3 = a_1 \\ \delta_{2m} a_m &= a_2 \\ \delta_{3m} a_m &= a_3 \end{aligned} \tag{1.14}$$

or

$$\delta_{im} a_m = a_i \tag{1.15}$$

3. In general,
$$\delta_{im} T_{mj} = T_{ij} \tag{1.16}$$

Remember, Eq. (1.16) represents nine equations.

Note: In Eq. (1.15), a_i are components of a vector (e.g., displacement, traction), and in Eq. (1.16), T_{ij} are components of a tensor (e.g., stress, strain).

4. The Kronecker delta, δ_{ij}, can be used to contract indices as follows:

$$\begin{aligned} \delta_{im}\delta_{mj} &= \delta_{ij} \\ \delta_{im}\delta_{mj}\delta_{jn} &= \delta_{in} \end{aligned} \tag{1.17}$$

5. If $\vec{e}_1, \vec{e}_2, \vec{e}_3$ are unit vectors perpendicular to each other,

$$\vec{e}_i \cdot \vec{e}_j = \delta_{ij} \tag{1.18}$$

or, $\vec{e}_1 \cdot \vec{e}_1 = \vec{e}_2 \cdot \vec{e}_2 = \vec{e}_3 \cdot \vec{e}_3 = 1$ and $\vec{e}_1 \cdot \vec{e}_2 = \vec{e}_1 \cdot \vec{e}_3 = \vec{e}_2 \cdot \vec{e}_1 = \ldots = 0$.

1.1.4 Manipulations

1. *Substitution.* Given two expressions,

$$a_i = U_{im} b_m \tag{1.19}$$

and

$$b_i = V_{im} c_m \tag{1.20}$$

write Eq. (1.19) in terms of U, V, and c.

To substitute Eq. (1.20) into Eq. (1.19), we need to change i in Eq. (1.20) to m. However, should we quit there, the new expression, $b_m = V_{mm} c_m$, would be meaningless, as the dummy index is repeated more than twice. Hence, we need a new dummy index, accomplished by changing m to n. Thus, Eq. (1.20) becomes

$$b_m = V_{mn} c_n \tag{1.21}$$

which, upon substituting Eq. (1.21) into Eq. (1.19), yields

$$a_i = U_{im} V_{mn} c_n \tag{1.22}$$

Expression (1.22) represents three equations (one free index) with nine terms on each right-hand side (two dummy indices).

2. *Multiplication.* If $p = a_m b_m$ and $q = c_m d_m$, then changing the dummy index m to n in the expression for q results in

$$pq = a_m b_m c_n d_n \tag{1.23}$$

3. *Vector dot product*. Given two vectors $\vec{a} = a_i\vec{e}_i$ and $b = b_i\vec{e}_i$, $\vec{a} \cdot \vec{b} = (a_i\vec{e}_i) \cdot (b_j\vec{e}_j) = a_i b_j(\vec{e}_i \cdot \vec{e}_j)$. Then, from Eq. (1.18), $\vec{a} \cdot \vec{b} = a_i b_j \delta_{ij} = a_i b_i$. Thus,

$$\vec{a} \cdot \vec{b} = a_i b_i \tag{1.24}$$

Using the dot product, we can also define the magnitude of a vector as

$$\left|\vec{a}\right| = \sqrt{\vec{a} \cdot \vec{a}} = \sqrt{a_i a_i} \tag{1.25}$$

The dot product of $\vec{a}$ and $\vec{b}$ is also given by $\vec{a} \cdot \vec{b} = |\vec{a}||\vec{b}|\cos\theta$, where θ is the interior angle between the two vectors. Furthermore, the dot product is commutative and distributive such that $\vec{a} \cdot \vec{b} = \vec{b} \cdot \vec{a}$ and $\vec{a} \cdot (\vec{b} + \vec{c}) = \vec{a} \cdot \vec{b} + \vec{a} \cdot \vec{c}$, respectively.

4. *Factoring*. If

$$T_{ij} n_j - \lambda n_i = 0 \tag{1.26}$$

then, using Eq. (1.15), $n_i = \delta_{ij} n_j$. So, Eq. (1.26) becomes

$$T_{ij} n_j - \lambda \delta_{ij} n_j = \left(T_{ij} - \lambda \delta_{ij}\right) n_j = 0 \tag{1.27}$$

1.2 TENSORS

1.2.1 Tensor Definition

A tensor is a linear transformation, denoted by $\underset{\sim}{T}$. It transforms any vector into another vector. A second-order tensor represents nine scalar quantities

$$\underset{\sim}{T}\vec{a} = \vec{b} \tag{1.28}$$

$\underset{\sim}{T}$ has the following properties

$$\begin{aligned} \underset{\sim}{T}\left(\vec{a} + \vec{b}\right) &= \underset{\sim}{T}\vec{a} + \underset{\sim}{T}\vec{b} \\ \underset{\sim}{T}(\alpha\vec{a}) &= \alpha\underset{\sim}{T}\vec{a} \end{aligned} \tag{1.29}$$

where a is a scalar.

1.2.2 Components of a Tensor

Let $\vec{e}_1, \vec{e}_2, \vec{e}_3$ be unit vectors in a rectangular Cartesian coordinate system. The Cartesian components of a vector $\vec{a}$ are given by $a_1 = \vec{a} \cdot \vec{e}_1$, $a_2 = \vec{a} \cdot \vec{e}_2$, and $a_3 = \vec{a} \cdot \vec{e}_3$, or

$$a_i = \vec{a} \cdot \vec{e}_i \tag{1.30}$$

Equivalently,

$$\vec{a} = a_1\vec{e}_1 + a_2\vec{e}_2 + a_3\vec{e}_3 = a_i\vec{e}_i \tag{1.31}$$

Now consider a tensor $\underset{\sim}{T}$. For any vector $\vec{a}$, $\vec{b} = \underset{\sim}{T}\vec{a}$ is a vector given by

$$\vec{b} = \underset{\sim}{T}\vec{a} = \underset{\sim}{T}(a_i\vec{e}_i) = a_i\underset{\sim}{T}\vec{e}_i \tag{1.32}$$

i.e., $\vec{b} = a_1\underset{\sim}{T}\vec{e}_1 + a_2\underset{\sim}{T}\vec{e}_2 + a_3\underset{\sim}{T}\vec{e}_3$. Note that $\underset{\sim}{T}\vec{e}_i$ are not perpendicular unit vectors. Using Eqs. (1.30) and (1.32), the components of $\vec{b}$ are

$$b_1 = \vec{e}_1 \cdot \vec{b} = \vec{e}_1 \cdot (a_i\underset{\sim}{T}\vec{e}_i) = a_i\vec{e}_1 \cdot \underset{\sim}{T}\vec{e}_i = a_1\vec{e}_1 \cdot \underset{\sim}{T}\vec{e}_1 + a_2\vec{e}_1 \cdot \underset{\sim}{T}\vec{e}_2 + a_3\vec{e}_2 \cdot \underset{\sim}{T}\vec{e}_3$$
$$b_2 = a_i\vec{e}_2 \cdot \underset{\sim}{T}\vec{e}_i$$
$$b_3 = a_i\vec{e}_3 \cdot \underset{\sim}{T}\vec{e}_i$$

or

$$b_i = a_j\vec{e}_i \cdot \underset{\sim}{T}\vec{e}_j \tag{1.33}$$

Now, $\vec{b} = \underset{\sim}{T}\vec{a}$ in matrix format is

$$\begin{bmatrix} b_1 \\ b_2 \\ b_3 \end{bmatrix} = \begin{bmatrix} T_{11} & T_{12} & T_{13} \\ T_{21} & T_{22} & T_{23} \\ T_{31} & T_{32} & T_{33} \end{bmatrix} \begin{bmatrix} a_1 \\ a_2 \\ a_3 \end{bmatrix}$$

from which we identify

$$b_i = T_{ij}a_j \tag{1.34}$$

Comparing Eqs. (1.33) and (1.34), we find that the components of $\underset{\sim}{T}$ are

$$T_{ij} = \vec{e}_i \cdot \underset{\sim}{T}\vec{e}_j \tag{1.35}$$

Demonstration.
What is the vector $\underset{\sim}{T}\vec{e}_j$?
Solution.

$$\underset{\sim}{T}\vec{e}_j = T_{kj}\vec{e}_k \tag{1.36}$$

To understand this solution, let us consider $\underset{\sim}{T}\vec{e}_2$ as an example.

$$\underset{\sim}{T}\vec{e}_2 = \begin{bmatrix} T_{11} & T_{12} & T_{13} \\ T_{21} & T_{22} & T_{23} \\ T_{31} & T_{32} & T_{33} \end{bmatrix} \begin{bmatrix} 0 \\ 1 \\ 0 \end{bmatrix} = \begin{bmatrix} T_{12} \\ T_{22} \\ T_{32} \end{bmatrix} \tag{1.37}$$

Thus, the j in $\underset{\sim}{T}\vec{e}_j$ determines the column, the same way that j denotes the column in the expression T_{ij}. By definition,

$$\begin{bmatrix} T_{12} \\ T_{22} \\ T_{32} \end{bmatrix} = T_{12}\vec{e}_1 + T_{22}\vec{e}_2 + T_{32}\vec{e}_3 = T_{k2}\vec{e}_k \tag{1.38}$$

Comparing Eqs. (1.37) and (1.38), we see $\underset{\sim}{T}\vec{e}_2 = T_{k2}\vec{e}_k$.

Now, refer to Eq. (1.35) with $i = 3$ and $j = 2$. From Eq. (1.37) and the definition of the dot product, Eq. (1.24),

$$T_{32} = \vec{e}_3 \cdot \underset{\sim}{T}\vec{e}_2 = \begin{bmatrix} T_{12} & T_{22} & T_{32} \end{bmatrix} \cdot \begin{bmatrix} 0 \\ 0 \\ 1 \end{bmatrix} = (0)(T_{12}) + (0)(T_{22}) + (1)(T_{32}) = T_{32} \tag{1.39}$$

Thus, the i in $\vec{e}_i$ in Eq. (1.35) determines the row, the same way that i denotes the row in the expression T_{ij}.

Note: The expression $\underset{\sim}{T}\vec{e}_j = T_{kj}\vec{e}_k$ will become useful when trying to determine the components of a tensor when the resulting transformation of original unit vectors is known.

1.2.3 Sum and Product of Tensors

Let $\underset{\sim}{T}$ and $\underset{\sim}{S}$ be two tensors. Then,

$$(\underset{\sim}{T} + \underset{\sim}{S})\vec{a} = \underset{\sim}{T}\vec{a} + \underset{\sim}{S}\vec{a} \tag{1.40}$$

Examining the components of the sum,

$$(\underset{\sim}{T} + \underset{\sim}{S})_{ij} = T_{ij} + S_{ij} \tag{1.41}$$

or, in matrix notation,

$$[\underset{\sim}{T} + \underset{\sim}{S}] = [\underset{\sim}{T}] + [\underset{\sim}{S}] \tag{1.42}$$

where $[\underset{\sim}{T}]$ is the matrix representation of the tensor $\underset{\sim}{T}$.

For the product of two tensors,

$$(\underset{\sim}{T}\underset{\sim}{S})\vec{a} = \underset{\sim}{T}(\underset{\sim}{S}\vec{a}) \tag{1.43}$$

$$(\underset{\sim}{T}\underset{\sim}{S})_{ij} = T_{im}S_{mj} \tag{1.44}$$

$$[\underset{\sim}{T}\underset{\sim}{S}] = [\underset{\sim}{T}][\underset{\sim}{S}] \tag{1.45}$$

Note: In general, $\underset{\sim}{T}\underset{\sim}{S} \neq \underset{\sim}{S}\,\underset{\sim}{T}$.

1.2.4 Identity Tensor

The identity tensor, $\underset{\sim}{I}$, is the linear transformation that transforms any vector into itself. Thus,

$$\underset{\sim}{I}\,\vec{a} = \vec{a} \tag{1.46}$$

from which it is evident that $\underset{\sim}{I}$ can be represented as

$$\underset{\sim}{I} = \delta_{ij} \tag{1.47}$$

1.2.5 Transpose of a Tensor

The transpose of a tensor $\underset{\sim}{T}$ is defined as the tensor $\underset{\sim}{T}^T$ that satisfies

$$\vec{a} \cdot (\underset{\sim}{T}\vec{b}) = \vec{b} \cdot (\underset{\sim}{T}^T\vec{a}) \tag{1.48}$$

Considering the unit vectors, for which Eq. (1.48) must also hold, $\vec{e}_i \cdot (\underset{\sim}{T}\vec{e}_j) = \vec{e}_j \cdot (\underset{\sim}{T}^T\vec{e}_i)$ which shows that

$$T_{ij} = \left(T^T\right)_{ji} \tag{1.49}$$

Note: The transpose of the product of two tensors is the product of the two tensors transposed in reverse order.

$$(\underset{\sim}{T}\underset{\sim}{S})^T = \underset{\sim}{S}^T\underset{\sim}{T}^T \tag{1.50}$$

1.2.6 Orthogonal Tensor

An orthogonal tensor, $\underset{\sim}{Q}$, is a linear transformation for which transformed vectors preserve their lengths and angles. Thus, given an orthogonal tensor $\underset{\sim}{Q}$,

$$|\underset{\sim}{Q}\vec{a}| = |\vec{a}| \tag{1.51}$$

Note: In general, $\underset{\sim}{Q}\vec{a} \neq \vec{a}$. Only their lengths are equal. Furthermore,

$$\cos\left(\vec{a}, \vec{b}\right) = \cos\left(\underset{\sim}{Q}\vec{a}, \underset{\sim}{Q}\vec{b}\right) \tag{1.52}$$

and, so it follows,

$$\left(\underset{\sim}{Q}\vec{a}\right) \cdot \left(\underset{\sim}{Q}\vec{b}\right) = \vec{a} \cdot \vec{b} \tag{1.53}$$

Note: In general, $\underset{\sim}{T}\,\underset{\sim}{T}^{T} \neq \underset{\sim}{T}^{T}\underset{\sim}{T} \neq \underset{\sim}{I}$, but $\underset{\sim}{Q}\,\underset{\sim}{Q}^{T} = \underset{\sim}{Q}^{T}\,\underset{\sim}{Q} = \underset{\sim}{I}$. Why? Let us take a look. From Eq. (1.48),

$$\left(\underset{\sim}{Q}\vec{a}\right) \cdot \left(\underset{\sim}{Q}\vec{b}\right) = \vec{b} \cdot \left(\underset{\sim}{Q}^{T}(\underset{\sim}{Q}\vec{a})\right) \tag{1.54}$$

Substituting Eq. (1.53) into Eq. (1.54),

$$\vec{a} \cdot \vec{b} = \vec{b} \cdot \left(\underset{\sim}{Q}^{T}(\underset{\sim}{Q}\vec{a})\right)$$

$$\vec{b} \cdot \vec{a} - \vec{b} \cdot \underset{\sim}{Q}^{T}\left(\underset{\sim}{Q}\vec{a}\right) = 0$$

$$\vec{b} \cdot \left(\underset{\sim}{I} - \underset{\sim}{Q}^{T}\,\underset{\sim}{Q}\right)\vec{a} = 0$$

Therefore, for arbitrary (nonzero) vectors $\vec{a}$ and $\vec{b}$,

$$\underset{\sim}{I} = \underset{\sim}{Q}^{T}\underset{\sim}{Q} = \underset{\sim}{Q}\,\underset{\sim}{Q}^{T} \tag{1.55}$$

Hence, for an orthogonal tensor $\underset{\sim}{Q}$, it is also true that

$$\underset{\sim}{Q} = \underset{\sim}{Q}^{-1} \tag{1.56}$$

In indicial notation

$$Q_{im}\,Q_{jm} = Q_{mi}\,Q_{mj} = \delta_{ij} \tag{1.57}$$

Note: The determinant of an orthogonal tensor is ±1, where

$$\det(\underset{\sim}{Q}) = \begin{cases} +1 & \text{indicates a rotation} \\ -1 & \text{indicates a reflection} \end{cases} \tag{1.58}$$

Demonstration.

A rigid body is rotated 90° by the right-hand rule about the $\vec{e}_3$ axis (see Figure 1.1).

1. Find the matrix representation of the tensor $\underset{\sim}{R}$ describing this rotation.
2. Examine whether $\underset{\sim}{R}$ is orthogonal.
3. Find the determinant of $\underset{\sim}{R}$.
4. Suppose this rigid body experiences a 90° right-hand rotation about the original $\vec{e}_1$ axis by the right-hand rule. Find the matrix representation of the tensor $\underset{\sim}{S}$ describing this rotation.
5. Find the final position of a point, p, originally at (1, 1, 0) after these two rotations.

Solution.

1. $\underset{\sim}{R}$ is the transformation

$$\begin{aligned} \underset{\sim}{R}\,\vec{e}_1 &= \vec{e}_2 \\ \underset{\sim}{R}\,\vec{e}_2 &= -\vec{e}_1 \\ \underset{\sim}{R}\,\vec{e}_3 &= \vec{e}_3 \end{aligned} \tag{1.59}$$

Recall Eq. (1.36).

$\underset{\sim}{R} = \begin{bmatrix} R_{11} & R_{12} & R_{13} \\ R_{21} & R_{22} & R_{23} \\ R_{31} & R_{32} & R_{33} \end{bmatrix}$ with $\underset{\sim}{R}\vec{e}_1$, $\underset{\sim}{R}\vec{e}_2$, and $\underset{\sim}{R}\vec{e}_3$ as columns. From Eqs. (1.36) and (1.59)

$$\begin{aligned} \underset{\sim}{R}\,\vec{e}_1 &= R_{11}\vec{e}_1 + R_{21}\vec{e}_2 + R_{31}\vec{e}_3 = \vec{e}_2 \\ \underset{\sim}{R}\,\vec{e}_2 &= R_{12}\vec{e}_1 + R_{22}\vec{e}_2 + R_{32}\vec{e}_3 = -\vec{e}_1 \\ \underset{\sim}{R}\,\vec{e}_3 &= R_{13}\vec{e}_1 + R_{23}\vec{e}_2 + R_{33}\vec{e}_3 = \vec{e}_3 \end{aligned} \tag{1.60}$$

So,

$$[\underset{\sim}{R}] = \begin{bmatrix} 0 & -1 & 0 \\ 1 & 0 & 0 \\ 0 & 0 & 1 \end{bmatrix}$$

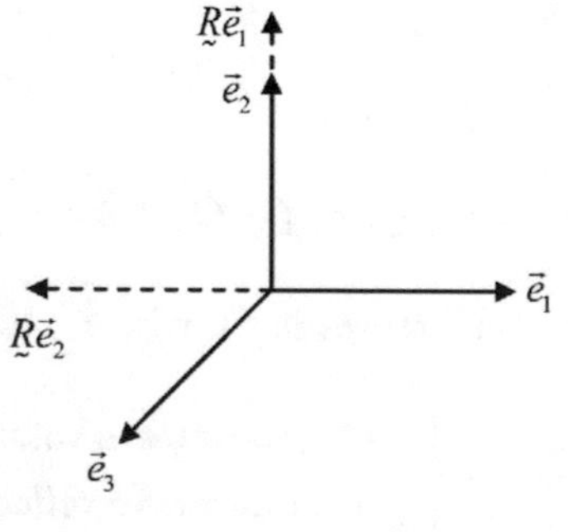

FIGURE 1.1: Standard coordinate system rotated 90° about the $\vec{e}_3$ axis.

2. Recall Eq. (1.55), $\underset{\sim}{R}$ is orthogonal iff ("if and only if") $\underset{\sim}{R}\underset{\sim}{R}^T = \underset{\sim}{I}$.

$$\left[\underset{\sim}{R}^T\right] = \begin{bmatrix} 0 & 1 & 0 \\ -1 & 0 & 0 \\ 0 & 0 & 1 \end{bmatrix}. \text{ So, } \underset{\sim}{R}\underset{\sim}{R}^T = \begin{bmatrix} 0 & -1 & 0 \\ 1 & 0 & 0 \\ 0 & 0 & 1 \end{bmatrix} \begin{bmatrix} 0 & 1 & 0 \\ -1 & 0 & 0 \\ 0 & 0 & 1 \end{bmatrix} = \begin{bmatrix} 1 & 0 & 0 \\ 0 & 1 & 0 \\ 0 & 0 & 1 \end{bmatrix} = \underset{\sim}{I}$$

Therefore, $\underset{\sim}{R}$ is orthogonal.

3. $\det(\underset{\sim}{R}) = \begin{vmatrix} 0 & -1 & 0 \\ 1 & 0 & 0 \\ 0 & 0 & 1 \end{vmatrix} = -(-1)(1) = 1$. Therefore, $\underset{\sim}{R}$ is a rotation.

4. $\underset{\sim}{S}\vec{e}_1 = \vec{e}_1$

   $\underset{\sim}{S}\vec{e}_2 = \vec{e}_3$

   $\underset{\sim}{S}\vec{e}_3 = -\vec{e}_2$

$\underset{\sim}{S}$ describes only the second rotation, not the combined total transformation. Following Eq. (1.60),

$$[\underset{\sim}{S}] = \begin{bmatrix} 1 & 0 & 0 \\ 0 & 0 & -1 \\ 0 & 1 & 0 \end{bmatrix}$$

5. The vector describing the location of point p is $\vec{r} = (1, 1, 0)^T$. Hence, the new location, p', of point p after both rotations is

$$\vec{r}\,' = \underset{\sim}{S}\underset{\sim}{R}r = \begin{bmatrix} 1 & 0 & 0 \\ 0 & 0 & -1 \\ 0 & 1 & 0 \end{bmatrix} \begin{bmatrix} 0 & -1 & 0 \\ 1 & 0 & 0 \\ 0 & 0 & 1 \end{bmatrix} \begin{bmatrix} 1 \\ 1 \\ 0 \end{bmatrix} = \begin{bmatrix} -1 \\ 0 \\ 1 \end{bmatrix}$$

So, $p' = (-1, 0, 1)$.

Note: The order of rotations matters, e.g., $\underset{\sim}{R}\underset{\sim}{S}\vec{r}$ yields $p' = (-1, 1, 0)$. This exemplifies the fact that given two tensors $\underset{\sim}{T}$ and $\underset{\sim}{S}$, $\underset{\sim}{T}\underset{\sim}{S} \neq \underset{\sim}{S}\underset{\sim}{T}$.

1.3 TENSOR SYMMETRY, PRINCIPAL VALUES, AND PRINCIPAL DIRECTIONS

1.3.1 Symmetric vs. Antisymmetric Tensor

A tensor, $\underset{\sim}{T}^{\text{symm}}$, is defined as symmetric iff $\underset{\sim}{T}^{\text{symm}} = \left(\underset{\sim}{T}^{\text{symm}}\right)^T$. This is different from orthogonal tensors. A symmetric tensor does not necessarily preserve lengths and angles. The components of a symmetric tensor satisfy

$$T_{ij} = T_{ji} \tag{1.61}$$

This means that $T_{12} = T_{21}$, $T_{31} = T_{13}$, and $T_{23} = T_{32}$. Also, the diagonal elements of $\underset{\sim}{T}^{\text{symm}}$ and $(\underset{\sim}{T}^{\text{symm}})^T$ must be equal.

A tensor, $\underset{\sim}{T}^{\text{asymm}}$, is antisymmetric iff $\underset{\sim}{T}^{\text{asymm}} = -(\underset{\sim}{T}^{\text{asymm}})^T$. The components of antisymmetric tensors satisfy

$$T_{ij} = -T_{ji} \tag{1.62}$$

This means that $T_{12} = -T_{21}$, $T_{31} = -T_{13}$, and $T_{23} = -T_{32}$. Also, the diagonal elements of $\underset{\sim}{T}^{\text{asymm}}$ must be zero. Antisymmetric tensors are also known as asymmetric or skew tensors.

Any tensor $\underset{\sim}{T}$ can always be decomposed into the sum of a symmetric and antisymmetric tensor.

$$\underset{\sim}{T} = \underset{\sim}{T}^{\text{symm}} + \underset{\sim}{T}^{\text{asymm}} \tag{1.63}$$

where

$$\begin{aligned} \underset{\sim}{T}^{\text{symm}} &= \frac{\underset{\sim}{T} + \underset{\sim}{T}^T}{2} \\ \underset{\sim}{T}^{\text{asymm}} &= \frac{\underset{\sim}{T} - \underset{\sim}{T}^T}{2} \end{aligned} \tag{1.64}$$

Another relevant decomposition of tensors is into spherical and deviatoric tensors (see Problem 14).

1.3.2 Eigenvalues and Eigenvectors

Consider a tensor $\underset{\sim}{T}$ and a vector $\vec{a}$. $\vec{a}$ is defined as an eigenvector of $\underset{\sim}{T}$ if it transforms under $\underset{\sim}{T}$ into a vector parallel to itself. This means

$$\underset{\sim}{T}\vec{a} = \lambda \vec{a} \tag{1.65}$$

where λ is called the eigenvalue. For definiteness, all eigenvectors will be of unit length.
Note: Any vector is an eigenvector of $\underset{\sim}{I}$, as $\underset{\sim}{I}\vec{a} = \vec{a}$ and $\lambda = 1$.

Now let $\vec{n}$ be a unit eigenvector. From Eqs. (1.65) and (1.46),

$$\underset{\sim}{T}\vec{n} = \lambda\vec{n} = \lambda\underset{\sim}{I}\vec{n} \tag{1.66}$$

with $\vec{n} \cdot \vec{n} = 1$. This implies

$$(\underset{\sim}{T} - \lambda\underset{\sim}{I})\vec{n} = 0 \tag{1.67}$$

or, in component form (let $\vec{n} = a_i \vec{e}_i$), using Eq. (1.27)

$$(T_{ij} - \lambda \delta_{ij}) a_j = 0 \tag{1.68}$$

with $a_j a_j = 1$. In long form,

$$\begin{aligned} (T_{11} - \lambda) a_1 + T_{12} a_2 + T_{13} a_3 &= 0 \\ T_{21} a_1 + (T_{22} - \lambda) a_2 + T_{23} a_3 &= 0 \\ T_{31} a_1 + T_{32} a_2 + (T_{33} - \lambda) a_3 &= 0 \end{aligned} \tag{1.69}$$

and

$$a_1^2 + a_2^2 + a_3^2 = 1 \tag{1.70}$$

Eqs. (1.68), (1.69), and (1.70) are used to solve for eigenvectors of a tensor. Eqs. (1.69) are a system of linear homogeneous equations in a_1, a_2, and a_3. Recall from linear algebra that the homogeneous equation $\underset{\sim}{A}\vec{x} = 0$ has only the trivial solution $\vec{x} = (0, 0, 0)$ unless $\underset{\sim}{A}$ is a singular matrix, i.e., $\underset{\sim}{A}^{-1}$ does not exist and det $(\underset{\sim}{A}) = 0$. Eq. (1.68) or (1.69) is of the form $\underset{\sim}{A}\vec{x} = 0$, and, in order to find nonzero eigenvectors, we must have det $(\underset{\sim}{A}) = 0$. Thus,

$$\left| \underset{\sim}{T} - \lambda \underset{\sim}{I} \right| = 0 \tag{1.71}$$

or, in long form,

$$\begin{vmatrix} T_{11} - \lambda & T_{12} & T_{13} \\ T_{21} & T_{22} - \lambda & T_{23} \\ T_{31} & T_{32} & T_{33} - \lambda \end{vmatrix} = 0 \tag{1.72}$$

Eq. (1.72) is a cubic equation in λ. It is known as the *characteristic equation* of $\underset{\sim}{T}$.

Demonstrations.

1. Given $\underset{\sim}{T}$ with $T_{21} = T_{31} = 0$, show that $\vec{e}_1$ is an eigenvector of $\underset{\sim}{T}$ with T_{11} as the corresponding eigenvalue.

Solution. Recall Eq. (1.65) that says $\underset{\sim}{T}\vec{a} = \lambda \vec{a}$. Also, recall Eq. (1.36). Then,
$\underset{\sim}{T}\vec{e}_1 = T_{11}\vec{e}_1 + T_{21}\vec{e}_2 + T_{31}\vec{e}_3 \Rightarrow \underset{\sim}{T}\vec{e}_1 = T_{11}\vec{e}_1$, which satisfies Eq. (1.65).

2. Given $\left[\underset{\sim}{T}\right] = \begin{bmatrix} 2 & 0 & 0 \\ 0 & 3 & 4 \\ 0 & 4 & -3 \end{bmatrix}$, find the eigenvalues, λ's, and corresponding eigenvectors, $\vec{n}$'s.

Solution. To find the eigenvalues, we will make use of the characteristic equation, Eq. (1.72). For the given $\underset{\sim}{T}$, Eq. (1.72) is

$$\begin{vmatrix} 2-\lambda & 0 & 0 \\ 0 & 3-\lambda & 4 \\ 0 & 4 & -3-\lambda \end{vmatrix} = 0 \tag{1.73}$$

which becomes

$$(2-\lambda)\,[(3-\lambda)\,(-3-\lambda)-16]=0$$

or

$$(2-\lambda)\,(\lambda^2-25)=0 \tag{1.74}$$

yielding $\lambda_1 = 2$, $\lambda_2 = 5$, and $\lambda_3 = -5$. To find the eigenvectors, we need to solve Eq. (1.68) for each λ. Eq. (1.68) for the given $\underset{\sim}{T}$ is

$$\begin{aligned} (2-\lambda)\,a_1 + 0a_2 + 0a_3 &= 0 \\ 0a_1 + (3-\lambda)\,a_2 + 4a_3 &= 0 \\ 0a_1 + 4a_2 + (-3-\lambda)\,a_3 &= 0 \end{aligned} \tag{1.75}$$

So, for $\lambda_1 = 2$, Eqs. (1.75) become

$$\begin{aligned} 0a_1 + 0a_2 + 0a_3 &= 0 \\ 0a_1 + 1a_2 + 4a_3 &= 0 \\ 0a_1 + 4a_2 + -5a_3 &= 0 \end{aligned}$$

These are two equations with two unknowns, which when solved give, $a_2 = a_3 = 0$ and a_1 unspecified. But, eigenvectors are of unit length. Using Eq. (1.70), $a_1^2 + a_2^2 + a_3^2 = 1$, so $a_1 = \pm 1$. Thus, the eigenvector, $\vec{n}_1$, corresponding to $\lambda_1 = 2$ is $\vec{n}_1 = \pm\, \vec{e}_1$.

For $\lambda_2 = 5$, Eqs. (1.75) become

$$\begin{aligned} -3a_1 + 0a_2 + 0a_3 &= 0 \\ 0a_1 + -2a_2 + 4a_3 &= 0 \\ 0a_1 + 4a_2 + -8a_3 &= 0 \end{aligned}$$

which, when solved, give $a_1 = 0$ and $a_2 = 2a_3$. Using Eq. (1.70),

$$(2a_3)^2 + a_3^2 = 1 \quad \Rightarrow \quad \begin{aligned} a_3 &= \pm\frac{1}{\sqrt{5}} \\ a_2 &= \pm\frac{2}{\sqrt{5}} \end{aligned}$$

Therefore, the eigenvector, $\vec{n}_2$, corresponding to $\lambda_2 = 5$ is $\vec{n}_2 = \pm\frac{1}{\sqrt{5}}(2\vec{e}_2 + \vec{e}_3)$ Similarly, for $\lambda_3 = -5$, $\vec{n}_3 = \pm\frac{1}{\sqrt{5}}(-\vec{e}_2 + 2\vec{e}_3)$.

Why have we reviewed the "Eigen-stuff"? As we will see in Chapters 2 and 3, they are used to determine the directions and magnitudes of maximum and minimum stresses and strains.

1.3.3 Principal Values and Principal Directions

Continuum biomechanics deals with real symmetric tensors, such as the stress tensor, strain tensor, and rate of deformation tensor.

Theorem: The eigenvalues of any real symmetric tensor are real and called *principal values*. The real eigenvectors (at least three) of any real symmetric tensor are called *principal directions*.

Theorem: For a real symmetric tensor, there exist three principal directions that are mutually perpendicular.

Let $\vec{n}_1, \vec{n}_2, \vec{n}_3$ be the principal directions (eigenvectors) of a real symmetric tensor $\underset{\sim}{T}$. From Eqs. (1.35) and (1.65), using $\vec{n}_i$ as the basis,

$$\begin{aligned}
T_{11} &= \vec{n}_1 \cdot \underset{\sim}{T}\vec{n}_1 = \vec{n}_1 \cdot (\lambda_1 \vec{n}_1) = \lambda_1 \\
T_{22} &= \vec{n}_2 \cdot \underset{\sim}{T}\vec{n}_2 = \vec{n}_2 \cdot (\lambda_2 \vec{n}_2) = \lambda_2 \\
T_{33} &= \vec{n}_3 \cdot \underset{\sim}{T}\vec{n}_3 = \vec{n}_3 \cdot (\lambda_3 \vec{n}_3) = \lambda_3 \\
T_{12} &= \vec{n}_{12} \cdot \underset{\sim}{T}\vec{n}_1 = \vec{n}_1 \cdot (\lambda_3 \vec{n}_2) = 0
\end{aligned}$$

Similarly, $T_{13} = T_{21} = T_{23} = T_{31} = T_{32} = 0$.

Thus, the original matrix $[\underset{\sim}{T}]$ in the $\vec{e}_1, \vec{e}_2, \vec{e}_3$ coordinate system (Figure 1.2)

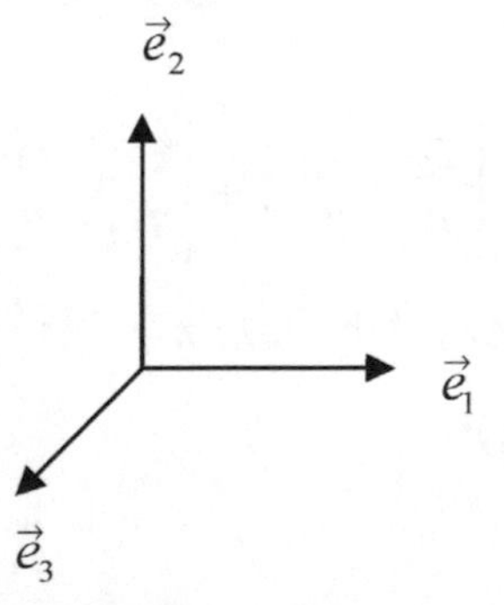

FIGURE 1.2: Reference coordinate system for $\underset{\sim}{T}$.

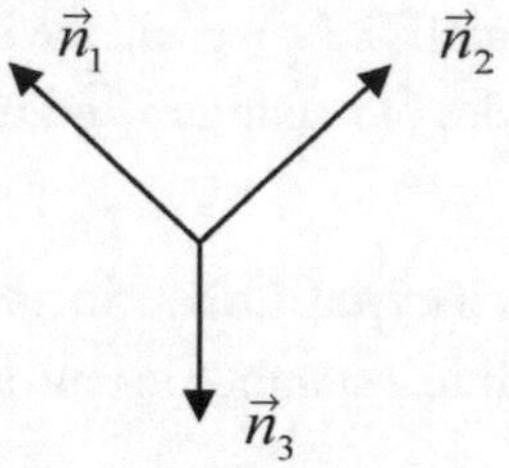

FIGURE 1.3: Rotated coordinate system in which $\underset{\sim}{T}$ is diagonal.

can be written in a rotated $\vec{n}_1, \vec{n}_2, \vec{n}_3$ coordinate system (Figure 1.3) as

$$\left[\underset{\sim}{T}\right]_{\vec{n}_1,\vec{n}_2,\vec{n}_3} = \begin{bmatrix} \lambda_1 & 0 & 0 \\ 0 & \lambda_2 & 0 \\ 0 & 0 & \lambda_3 \end{bmatrix} \tag{1.76}$$

which is a diagonal matrix, where the diagonal elements are the principal values (eigenvalues) of $[\underset{\sim}{T}]$.

1.4 OTHER USEFUL TENSOR RELATIONSHIPS

1.4.1 Scalar Invariants of a Tensor and the Cayley-Hamilton Theorem

The scalar characteristic equation of a tensor $\underset{\sim}{T}$, $\left|T_{ij} - \lambda\delta_{ij}\right| = 0$, is a cubic equation in λ,

$$\lambda^3 - I_1\lambda^2 + I_2\lambda - I_3 = 0 \tag{1.77}$$

where I_1, I_2, I_3 are invariants given by

$$I_1 = T_{11} + T_{22} + T_{33} = T_{ii} \tag{1.78}$$

$$\begin{aligned} I_2 &= \begin{vmatrix} T_{11} & T_{12} \\ T_{21} & T_{22} \end{vmatrix} + \begin{vmatrix} T_{22} & T_{23} \\ T_{32} & T_{33} \end{vmatrix} + \begin{vmatrix} T_{11} & T_{13} \\ T_{31} & T_{33} \end{vmatrix} \\ &= (T_{11}T_{22} - T_{21}T_{12}) + (T_{22}T_{33} - T_{32}T_{23}) + (T_{11}T_{33} - T_{31}T_{13}) \\ &= \frac{1}{2}(T_{ii}T_{jj} - T_{ij}T_{ij}) \end{aligned} \tag{1.79}$$

and

$$I_3 = \det(\underset{\sim}{T}) \tag{1.80}$$

Alternatively,

$$\begin{aligned} I_1 &= \lambda_1 + \lambda_2 + \lambda_3 \\ I_2 &= \lambda_1\lambda_2 + \lambda_2\lambda_3 + \lambda_3\lambda_1 \\ I_3 &= \lambda_1\lambda_2\lambda_3 \end{aligned} \tag{1.81}$$

The Cayley–Hamilton theorem is a useful result from linear algebra. It says that a square matrix satisfies its own characteristic equation. Thus, for the 3x3 tensors we have been considering,

$$\underset{\sim}{T}^3 - I_1\underset{\sim}{T}^2 + I_2\underset{\sim}{T} - I_3\underset{\sim}{I} = 0 \tag{1.82}$$

1.4.2 Trace of a Tensor

The trace of a tensor $\underset{\sim}{T}$, denoted as tr $(\underset{\sim}{T})$, is the sum of the diagonal terms.

$$\text{tr}\,(\underset{\sim}{T}) = T_{ii} = T_{11} + T_{22} + T_{33} \tag{1.83}$$

Also,

$$\text{tr}\,(\underset{\sim}{T}) = \text{tr}\,(\underset{\sim}{T}^T) \tag{1.84}$$

It is notable that the invariants given in section 1.4.1 can all be written in terms of tr $(\underset{\sim}{T})$ (see Problem 17).

1.4.3 Tensor-Valued Functions of a Scalar

Let $\underset{\sim}{T} = \underset{\sim}{T}(t)$ and $\underset{\sim}{S} = \underset{\sim}{S}(t)$ be tensor-valued functions of a scalar, t (e.g., time). Then,

$$\frac{\mathrm{d}}{\mathrm{d}t}(\underset{\sim}{T}\underset{\sim}{S}) = \frac{\mathrm{d}\underset{\sim}{T}}{\mathrm{d}t}\underset{\sim}{S} + \underset{\sim}{T}\frac{\mathrm{d}\underset{\sim}{S}}{\mathrm{d}t} \tag{1.85}$$

$$\frac{\mathrm{d}}{\mathrm{d}t}(\underset{\sim}{T} + \underset{\sim}{S}) = \frac{\mathrm{d}\underset{\sim}{T}}{\mathrm{d}t} + \frac{\mathrm{d}\underset{\sim}{S}}{\mathrm{d}t} \tag{1.86}$$

$$\frac{\mathrm{d}}{\mathrm{d}t}\left[\alpha(t)\underset{\sim}{T}\right] = \frac{\mathrm{d}\alpha}{\mathrm{d}t}\underset{\sim}{T} + \alpha\frac{\mathrm{d}\underset{\sim}{T}}{\mathrm{d}t} \tag{1.87}$$

$$\frac{\mathrm{d}}{\mathrm{d}t}(\underset{\sim}{T}\vec{a}) = \frac{\mathrm{d}\underset{\sim}{T}}{\mathrm{d}t}\vec{a} + \underset{\sim}{T}\frac{\mathrm{d}\vec{a}}{\mathrm{d}t} \tag{1.88}$$

$$\frac{\mathrm{d}}{\mathrm{d}t}(\underset{\sim}{T}^T) = \left(\frac{\mathrm{d}\underset{\sim}{T}}{\mathrm{d}t}\right)^T \tag{1.89}$$

In these expressions, α is a scalar and $\vec{a}$ is a vector.

1.4.4 Gradient and Divergence of Scalar, Vector, and Tensor Fields

Let $\phi\ (\vec{r})$, be a scalar-valued function of the position vector $\vec{r}$; i.e., $\phi(x_1, x_2, x_3)$. For each position $\vec{r}$, ϕ gives the value of a scalar (e.g., density, temperature, electric potential) at that point; i.e., $\phi\ (\vec{r})$ describes a scalar field. $\nabla\phi(\vec{r})$, the gradient of ϕ, is a vector field whose form in Cartesian coordinates is $\dfrac{\partial \phi}{\partial x_i}\vec{e}_i$.

$$\nabla \equiv \frac{\partial}{\partial x_i}\vec{e}_i = \begin{bmatrix} \partial/\partial x_1 \\ \partial/\partial x_2 \\ \partial/\partial x_3 \end{bmatrix} \tag{1.90}$$

So,

$$\nabla\phi = \frac{\partial \phi}{\partial x_1}\vec{e}_1 + \frac{\partial \phi}{\partial x_2}\vec{e}_2 + \frac{\partial \phi}{\partial x_3}\vec{e}_2 = \frac{\partial \phi}{\partial x_i}\vec{e}_i = \begin{bmatrix} \partial\phi/\partial x_1 \\ \partial\phi/\partial x_2 \\ \partial\phi/\partial x_3 \end{bmatrix} \tag{1.91}$$

Let $\vec{v}\ (\vec{r})$ be a vector-valued function of position (e.g., displacement or velocity fields). $\nabla\vec{v}$, the gradient of $\vec{v}$, is a second-order tensor. In Cartesian coordinates

$$\left(\nabla\vec{v}\right)_{ij} = \frac{\partial v_i}{\partial x_j} \tag{1.92}$$

or in matrix notation,

$$\left[\nabla\vec{v}\right] = \begin{bmatrix} \dfrac{\partial v_1}{\partial x_1} & \dfrac{\partial v_1}{\partial x_2} & \dfrac{\partial v_1}{\partial x_3} \\ \dfrac{\partial v_2}{\partial x_1} & \dfrac{\partial v_2}{\partial x_2} & \dfrac{\partial v_2}{\partial x_3} \\ \dfrac{\partial v_3}{\partial x_1} & \dfrac{\partial v_3}{\partial x_2} & \dfrac{\partial v_3}{\partial x_3} \end{bmatrix} \tag{1.93}$$

The divergence of $\vec{v}(\vec{r})$, div $\vec{v}(\vec{r})$, is defined to be a scalar field given by the trace of the gradient of $\vec{v}$.

$$\mathrm{div}(\vec{v}) \equiv \mathrm{tr}\,(\nabla\vec{v}) = \nabla\cdot\vec{v} \tag{1.94}$$

In Cartesian coordinates,

$$\mathrm{div}(\vec{v}) = \frac{\partial v_1}{\partial x_1} + \frac{\partial v_2}{\partial x_2} + \frac{\partial v_3}{\partial x_3} = \frac{\partial v_i}{\partial x_i} \tag{1.95}$$

Finally, let $\underset{\sim}{T}(\vec{r})$ be a tensor field. The divergence of a tensor field is a vector field. In Cartesian coordinates,

$$\nabla \cdot \underset{\sim}{T} = \left(\frac{\partial T_{im}}{\partial x_m}\right) \vec{e}_i \tag{1.96}$$

Note that the expressions for the gradient and divergence have different forms in cylindrical and spherical coordinates from those given for Cartesian coordinates.

Summary:

- Gradient increases rank
- scalar → vector
- vector → tensor

- Divergence reduces rank
- tensor → vector
- vector → scalar

1.5 PROBLEMS

1. Given $\left[\underset{\sim}{S}\right] = \begin{bmatrix} 7 & 0 & 11 \\ 4 & 2 & 1 \\ 0 & 6 & 2 \end{bmatrix}$

   Evaluate

   (a) S_{ii}

   (b) $S_{ij}S_{ij}$

   (c) $S_{jk}S_{jk}$

   (d) $S_{mn}S_{nm}$

2. Consider the expression $j_a = f_a + D_{ab}e_b$.

   (a) How many equations are described?

   (b) How many terms are on the right-hand side?

   (c) Which are the free indices?

   (d) Which are the dummy indices?

3. Consider the expression $T_{ij} = A_{ijmn}x_m x_n + B_{ip}C_{jp}$.

   (a) How many equations are described?

   (b) How many terms are on the right-hand side?

   (c) Which are the free indices?

   (d) Which are the dummy indices?

   (e) Now answer a)–d) again for $T_{ij} = A_{ijmn}x_m y_n + B_{iprt}E_{jt}c_p d_r$.

4. Consider the expression $T_{ijk} = A_{ijkmn} x_m x_n$.
(a) How many equations are described?
(b) How many terms are on the right-hand side?
(c) What are the free indices?
(d) What are the dummy indices?

5. Why can we say $a_i = \vec{a} \cdot \vec{e}_i$ and $\vec{a} = a_i \vec{e}_i$? In other words, why does this not violate basic algebraic rules?

6. Show that $g_{ijk} c_i c_j c_k = g_{111} + g_{121} + g_{211} + g_{221} + g_{112} + g_{122} + g_{212} + g_{222}$ for a particular value of c_1 and c_2. In this case, i, j, and k take on the values 1 and 2. What is this value?

7. Evaluate $\delta_{im} \delta_{mj} \delta_{jn} \delta_{nx}$.

8. Given that $T_{ij} = 2\mu E_{ij} + \lambda E_{kk} \delta_{ij}$, show that
(a) $W = \frac{1}{2} T_{ij} E_{ij} = \mu E_{ij} E_{ij} + \frac{1}{2} \lambda \left(E_{kk}\right)^2$
(b) $N = T_{ij} T_{ij} = 4\mu^2 E_{ij} E_{ij} + (4\mu\lambda + 3\lambda^2)\left(E_{kk}\right)^2$

9. Show that $\vec{a} \cdot \left(\underset{\sim}{T} \vec{b} \right) = \vec{b} \cdot \left(\underset{\sim}{T}^T \vec{a} \right) = T_{ij} a_i b_j$

10. Show that $\begin{bmatrix} 1/\sqrt{2} \\ 0 \\ -1/\sqrt{2} \end{bmatrix}$ is an eigenvector of the matrix $\begin{bmatrix} 5 & 13 & 0 \\ 6 & -17 & 6 \\ 0 & 67 & 5 \end{bmatrix}$, and find its corresponding eigenvalue.

11. Consider $\left[\underset{\sim}{T}\right] = \begin{bmatrix} 5 & 16 & 4 \\ 0 & 8 & 2 \\ 0 & 0 & -10 \end{bmatrix}$.

Find its scalar invariants and then evaluate the eigenvalues using $\lambda^3 - I_1\lambda^2 + I_2\lambda - I_3 = 0$. (You can use Matlab©'s "roots" command to solve the cubic equation.)

12. Find the eigenvalues and eigenvectors for the tensor $\underset{\sim}{T}$, where $\left[\underset{\sim}{T}\right] = \begin{bmatrix} 8 & 0 & 0 \\ 0 & 8 & 0 \\ 0 & 0 & 6 \end{bmatrix}$. (Hint: You will need three orthogonal eigenvectors.)

13. Given $\left[\underset{\sim}{T}\right] = \begin{bmatrix} 7 & 1 & 3 \\ 8 & 3 & 2 \\ 2 & 8 & 1 \end{bmatrix}$,

(a) Decompose the tensor into its symmetric and antisymmetric parts.
(b) Based on your answer to (a), provide a general statement about T_{ii} if a tensor is antisymmetric.

14. Eqs. (1.64) present the decomposition of a tensor into its symmetric and antisymmetric parts. Another important tensor decomposition is into *spherical* (or hydrostatic) and *deviatoric* parts. Given the following definitions for the spherical and deviatoric components,

$$\mathrm{sph}(\underset{\sim}{T}) = \frac{1}{3}\mathrm{tr}(\underset{\sim}{T})\underset{\sim}{I} \qquad \text{and} \qquad \mathrm{dev}(\underset{\sim}{T}) = \underset{\sim}{T} - \mathrm{sph}(\underset{\sim}{T}) \tag{1.97}$$

verify
(a) $\underset{\sim}{T} = \mathrm{sph}(\underset{\sim}{T}) + \mathrm{dev}(\underset{\sim}{T})$
(b) $\mathrm{sph}(\mathrm{dev}\,(\underset{\sim}{T}))=0$
(c) Is $\mathrm{sph}(\underset{\sim}{T})$ symmetric or antisymmetric? Describe why $\mathrm{dev}(\underset{\sim}{T})$ is symmetric only if $\underset{\sim}{T}$ is symmetric.
(d) Decompose $\underset{\sim}{T}$ from problem 13 into its spherical and deviatoric parts.
Note: This decomposition is helpful when modeling the behavior of incompressible materials.

15. $\underset{\sim}{T}$ transforms every vector into its mirror image with respect to the 1–2 plane (i.e., let $\vec{e}_3$ be perpendicular to the plane of reflection).
(a) Find the matrix representation of $\underset{\sim}{T}$.
(b) Verify that $\underset{\sim}{T}$ is indeed a tensor (or linear transformation).
(c) Prove that $\underset{\sim}{T}$ is an orthogonal tensor.
(d) Show that $\underset{\sim}{T}$ is a reflection by calculating its determinant.

16. A tensor, $\underset{\sim}{A}$, transforms a Cartesian coordinate system by rotating it 30° counterclockwise about the $\vec{e}_2$ axis and extending vectors along the $\vec{e}_2$ axis to three times their original length. You will find Figure 1.4 helpful.
(a) Find $\underset{\sim}{A}$.
(b) A plane intersects the original coordinate axes at (1, 0, 0), (0, 1, 0), and (0, 0, 1). The unit vector normal to this plane is transformed by $\underset{\sim}{A}$. Find the vector, $\vec{v}$, resulting from this transformation.
(c) Why is $\underset{\sim}{A}$ not an orthogonal tensor?

17. The invariants of a second-order tensor can all be expressed as a function of the trace of the tensor and the trace of the tensor squared or cubed. What is I_1 in terms of $\mathrm{tr}(\underset{\sim}{T})$? Also, show that I_2 and I_3 can be written as

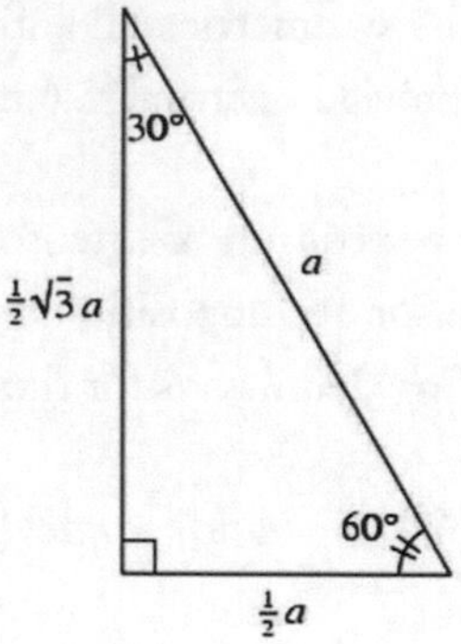

FIGURE 1.4: Relationships among the lengths of the legs of a 30°-60°-90° triangle.

$$I_2 = \frac{1}{2}\left[\left(\mathrm{tr}(\underset{\sim}{T})\right)^2 - \mathrm{tr}(\underset{\sim}{T}^2)\right]$$
$$I_3 = \frac{1}{3}\left[\mathrm{tr}(\underset{\sim}{T}^3) - \frac{3}{2}\mathrm{tr}(\underset{\sim}{T}^2)\mathrm{tr}(\underset{\sim}{T}) + \frac{1}{2}\left(\mathrm{tr}(\underset{\sim}{T})\right)^3\right] \quad (1.98)$$

(Hint: Use the Cayley-Hamilton theorem for I_3)

18. Show that the trace satisfies tr $(\alpha\underset{\sim}{S} + \beta\underset{\sim}{T}) = \alpha$ tr $(\underset{\sim}{S}) + \beta\mathrm{tr}(\underset{\sim}{T})$.

19. In this chapter, we described the dot product of two vectors in indicial notation. In this problem, we will examine the cross product of two vectors. Show that the cross product of two vectors is given by

$$\vec{a} \times \vec{b} = (\varepsilon_{ijk}\, a_j b_k)\, \vec{e}_i \quad (1.99)$$

where

$$\varepsilon_{ijk} = \vec{e}_i \cdot (\vec{e}_j \times \vec{e}_k), \quad (1.100)$$

or equivalently,

$$\varepsilon_{ijk} = \begin{cases} 1 & \text{when } ijk \text{ is an even permutation of 123 (i.e., 123, 312, 231)} \\ -1 & \text{when } ijk \text{ is an odd permutation of 123 (i.e., 321, 132, 213)} \\ 0 & \text{if any two of } ijk \text{ are equal (i.e., 113, 232, 333)} \end{cases} \quad (1.101)$$

Recall, the cross product of two vectors is $(a_2 b_3 - a_3 b_2)\,\vec{e}_1 - (a_1 b_3 - a_3 b_1)\,\vec{e}_2 + (a_1 b_2 - a_2 b_1)\,\vec{e}_3$.

Note: An alternate definition of the cross product is $\vec{a} \times \vec{b} = (ab \sin \theta)\vec{n}$, where a and b are magnitudes, θ is the angle between the vectors $\vec{a}$ and $\vec{b}$, and $\vec{n}$ is a unit vector perpendicular to both $\vec{a}$ and $\vec{b}$ chosen such that the set $(\vec{a}, \vec{b}, \vec{n})$ make a right-handed triad.

20. Show $\nabla \cdot (\alpha f \underset{\sim}{I}) = \alpha \nabla f$, where α is a constant and f is a scalar valued function. What type of field (scalar, vector, tensor) is the left-hand side? The right-hand side? It is important to remember that both sides of an equation must match in terms of *units* and *type of field.*

21. For each of the following, 1) indicate what the operation produces (i.e., scalar, vector, or tensor) and 2) express the operation with its result in indicial notation with respect to Cartesian coordinates.

(a) Gradient of a scalar
(b) Gradient of a vector
(c) Divergence of a vector
(d) Divergence of a tensor

22. Tensors can be formed from the outer product of two vectors. The outer product, or dyadic product, of two vectors is formed as $[\vec{a} \otimes \vec{b}]_{ij} = a_i b_j$. It also conforms to the following rules when forming inner products: $(\vec{a} \otimes \vec{b}) \cdot \vec{c} = \vec{a}(\vec{b} \cdot \vec{c})$ and $a\ (\vec{b} \otimes \vec{c}) = (\vec{a} \cdot \vec{b})\ \vec{c}$. Show that any tensor $\underset{\sim}{T}$ can be expressed as $\underset{\sim}{T} = T_{ij}\ (\vec{e}_i \otimes \vec{e}_j)$.

• • • •

CHAPTER 2

Kinematics of a Continuum

2.1 DESCRIPTION OF THE MOTION OF A CONTINUUM

Kinematics is the study of motion dealing with position, body (configuration), and time. Kinematic analyses are based on relationships among position, velocity, and acceleration. In this chapter, we will develop the basic kinematic concepts essential to continuum mechanics.

When we describe the flow of water in a river, we do not desire to identify the locations from which each particle of water originates. Also, we are not interested in describing the water's velocity, or other quantities, with respect to the origin or its reference position. Instead, we want to know the instantaneous velocity field (and other properties) and its evolution in time with respect to a current position of interest.

The Eulerian method analyzes what happens at every fixed point in space, whereas the Lagrangian method follows the trajectory of each individual particle. If we describe tensor quantities with respect to the origin, we use an Eulerian (spatial) description. If we use a particular position of the object at some time, having followed the individual particles during the motion, we are using the Lagrangian (material) description.

Let $O(x_1, x_2, x_3)$ be a fixed frame as shown in Figure 2.1. A body at time $t = t_0$ occupies a certain region of physical space. The position of a particle at $t = t_0$ is described by its position vector $\vec{a}$, measured from a fixed point, O. At a later time t, let the position vector of the particle be $\vec{x}$ (i.e., as time goes on, the particle moves). Thus,

$$\vec{x} = \vec{x}(\vec{a}, t) \tag{2.1}$$

with $\vec{x}(\vec{a}, t_0) = \vec{a}$, and where $\vec{a} = a_1\vec{e}_1 + a_2\vec{e}_2 + a_3\vec{e}_3$ and $\vec{x} = x_1\vec{e}_1 + x_2\vec{e}_2 + x_3\vec{e}_3$. Thus, Eq. (2.1) becomes

$$\begin{aligned} x_1 &= x_1(a_1, a_2, a_3, t) \\ x_2 &= x_2(a_1, a_2, a_3, t) \\ x_3 &= x_3(a_1, a_2, a_3, t) \end{aligned}$$

or

$$x_i = x_i(a_1, a_2, a_3, t) \tag{2.2}$$

with $a_i = x_i(a_1, a_2, a_3, t_0)$.

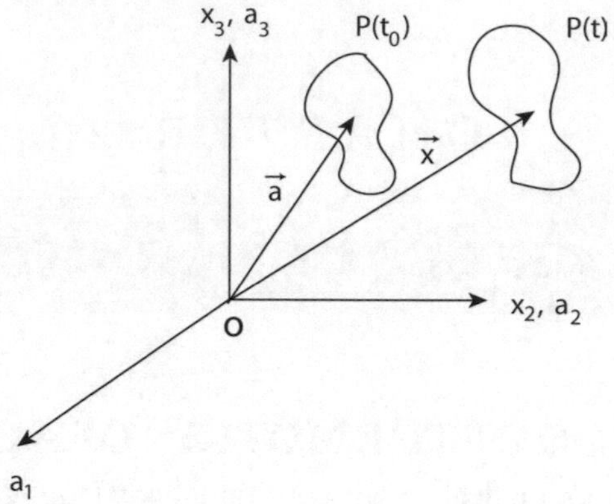

FIGURE 2.1: Motion of a continuum. At $t = t_0$, a particle is located at material coordinates a_1, a_2, a_3 defining the position vector $\vec{a}$. At a later time, t, motion has moved this particle to coordinates x_1, x_2, x_3 defining the position vector $\vec{x}$.

In Eq. (2.2), the triple set (a_1, a_2, a_3) identifies the different particles of the body and are said to be the *material coordinates*. Eq. (2.1) or (2.2) is said to define a motion for a continuum. For a specific particle, they define the path line or trajectory followed by that particle.

2.2 MATERIAL VS. SPATIAL DESCRIPTION

When a continuum is in motion, quantities that are associated with specific particles (temperature θ, velocity $\vec{v}$, etc.) change with time and, perhaps, position. We can depict these changes using Lagrangian (material) or Eulerian (spatial) descriptions.

1. *Lagrangian (Material) description*—material coordinates are the independent variables

$$\theta = \theta(a_1, a_2, a_3, t)$$

$$\vec{v} = \vec{v}(a_1, a_2, a_3, t)$$

Every particle is described by its coordinates at time t_0. a_1, a_2, a_3 are the material coordinates (i.e., we follow the particles and express θ, $\vec{v}$ as functions of the particles).

2. *Eulerian (Spatial) description*—spatial coordinates are the independent variables

$$\theta = \theta(x_1, x_2, x_3, t)$$

$$\vec{v} = \vec{v}(x_1, x_2, x_3, t)$$

x_1, x_2, x_3 are known as spatial coordinates (i.e., we observe the changes at fixed locations). Spatial positions are occupied by different particles at different times.

Note: The material and spatial descriptions are related by the motion.

Demonstration.
Consider the motion $\vec{x} = \vec{a} + kta_2\,\vec{e}_1$, shown in Figure 2.2, where the material coordinates (a_1, a_2, a_3) give the position of a particle at $t = 0$.

1. Sketch the configuration at time t' for a body which, at $t = 0$, has the shape of a cube of unit sides.
2. If the temperature is given by the spatial description, $\theta = x_1 + x_2$,
   (a) Find the material description of θ.
   (b) Obtain the velocity and rate of change of temperature for particular material particles. Express the answer in both material and spatial coordinates.

Solution.

1. Break down the equations of motion into

$$x_1 = a_1 + kta_2$$

$$x_2 = a_2$$

$$x_3 = a_3$$

At $t = 0$, the coordinates of point O are $(a_1, a_2, a_3)_O = (0,0,0)$. At $t = t'$, the coordinates of O are $(x_1, x_2, x_3)_O = (0,0,0)$, i.e., point O remains at (0,0,0).

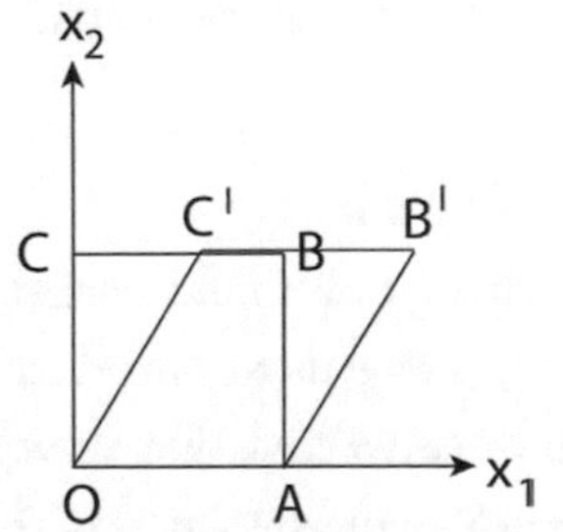

FIGURE 2.2: Unit cube subjected to the motion described by $\vec{x} = \vec{a} + kta_2\,\vec{e}_1$.

At $t = 0$, point A is at $(a_1, a_2, a_3)_A = (1,0,0)$. At $t = t'$, point A is at $(x_1, x_2, x_3)_A = (1,0,0)$. Thus, A also remains stationary. So, for any particle on OA, $(a_1, a_2, a_3)_{OA} = (a_1,0,0)$ at $t = 0$ and $(x_1, x_2, x_3)_{OA} = (a_1,0,0)$ at $t = t'$. Line OA is stationary.

Now, looking at the different lines, any particle on material line CB has position $(a_1, a_2, a_3)_{CB} = (a_1,1,0)$ at $t = 0$ and $(x_1, x_2, x_3)_{CB} = (a_1+kt',1,0)$ at $t = t'$. So, line CB is displaced to the right by distance kt'. Any particle on material line OC has position $(a_1, a_2, a_3)_{OC} = (0, a_2,0)$ at $t = 0$ and $(x_1, x_2, x_3)_{OC} = (a_2kt', a_2,0)$ at $t = t'$. So, every particle on line OC moves horizontally to the right through a distance linearly proportional to its height. Similarly, material line BA moves as line OC. This motion is known as *simple shearing motion.*

2. (a) $\theta = x_1 + x_2$ and $\vec{x} = \vec{a} + kta_2\vec{e}_1$. So, the material description of θ is

$$\theta = (a_1 + kta_2) + a_2 = a_1 + (kt + 1)\, a_2$$

(b) Because a particular material particle is designated by a specific a_i, the velocity of that particle is given by

$$v_i = \left.\frac{\partial x_i}{\partial t}\right|_{a_i \text{ fixed}}$$

So, $v_1 = ka_2$, $v_2 = v_3 = 0$ is the material description. Furthermore, as $x_2 = a_2$, the spatial description is $v_1 = kx_2$, $v_2 = v_3 = 0$. To calculate the rate of temperature change for a material particle,

$$\left.\frac{\partial \theta}{\partial t}\right|_{a_i \text{ fixed}} = \left.\frac{\partial}{\partial t}(x_1 + x_2)\right|_{a_i \text{ fixed}} = v_1 + v_2 = ka_2 = kx_2$$

Note: As $\theta = x_1 + x_2$, it is independent of time in the spatial description. However, each particle experiences temperature changes as it flows from one spatial coordinate to another.

2.3 MATERIAL DERIVATIVE

The concept just presented, namely, change of a field quantity independent of time due to "flow" from one spatial coordinate to another, allows us to introduce the *material derivative.* The material derivative (also called the Lagrangian or convective derivative) is defined as the time rate of change of a quantity (such as temperature or velocity) associated with a material particle. The material derivative is a derivative taken with respect to a coordinate system moving with velocity $\vec{v}$, and it is often used in fluid and classical mechanics. It is denoted by $\frac{\mathrm{D}}{\mathrm{D}t}$. When the material description of the quantity is used, i.e., $\theta = \theta\,(a_1, a_2, a_3, t)$, then

$$\frac{\mathrm{D}\theta}{\mathrm{D}t} = \left.\frac{\partial \theta}{\partial t}\right|_{a_i \text{ fixed}} \tag{2.3}$$

However, when the spatial description of the quantity is used, i.e., $\theta = \theta\,(x_1, x_2, x_3, t)$ with $x_i = x_i\,(a_1, a_2, a_3, t)$, then

$$\frac{\mathrm{D}\theta}{\mathrm{D}t} = \left.\frac{\partial \theta}{\partial t}\right|_{a_i \text{ fixed}} = \frac{\partial \theta}{\partial x_1}\frac{\partial x_1}{\partial t} + \frac{\partial \theta}{\partial x_2}\frac{\partial x_2}{\partial t} + \frac{\partial \theta}{\partial x_3}\frac{\partial x_3}{\partial t} + \left.\frac{\partial \theta}{\partial t}\right|_{x_i \text{ fixed}} \tag{2.4}$$

where the $\frac{\partial x_i}{\partial t}$ are obtained with fixed values of a_i. In Cartesian coordinates, $\left.\frac{\partial x_i}{\partial t}\right|_{a_i \text{ fixed}} = v_i$. Thus,

$$\frac{\mathrm{D}\theta}{\mathrm{D}t} = \frac{\partial \theta}{\partial t} + v_1\frac{\partial \theta}{\partial x_1} + v_2\frac{\partial \theta}{\partial x_2} + v_3\frac{\partial \theta}{\partial x_3} = \frac{\partial \theta}{\partial t} + v_i\frac{\partial \theta}{\partial x_i} = \frac{\partial \theta}{\partial t} + \vec{v} \cdot \nabla\theta \tag{2.5}$$

where the last equality has made use of Eqs. (1.24) and (1.91). Remember, in these equations, θ is a scalar valued function of stationary spatial coordinates, and $\vec{v}$ is a vector valued function of stationary spatial coordinates. Let us pause for a moment to consider what Eq. (2.5) means. Consider a particle at position x_i at time t that has with it an associated value (e.g., scalar or vector) $\theta(x_i,t)$. At a later time, $t + \mathrm{d}t$, the particle has moved to a new location, $x_i + v_i\,\mathrm{d}t$, and has associated with it the value $\theta(x_i + v_i\,\mathrm{d}t, t + \mathrm{d}t)$. The associated value θ has changed by an amount $\theta(x_i + v_i\,\mathrm{d}t, t + \mathrm{d}t) - \theta(x_i,t)$ due to both change in ti me and spatial location. The material derivative captures the nature of this change in the infinitesimal limit.

Demonstration. Consider the previous example with motion $\vec{x} = \vec{a} + kta_2\,\vec{e}_1$. The velocity was found to be $\vec{v} = kx_2\,\vec{e}_1$, and the temperature field was given as $\theta = x_1 + x_2$. Find the material derivative, $\frac{\mathrm{D}\theta}{\mathrm{D}t}$.

Solution.

$$\frac{D\theta}{Dt} = \frac{\partial \theta}{\partial t} + \vec{v} \cdot \nabla\theta, \text{ where } \nabla\theta = \frac{\partial \theta}{\partial x_i}\vec{e}_i = \vec{e}_1 + \vec{e}_2 \text{. So,}$$

$$\frac{D\theta}{Dt} = 0 + (kx_2\vec{e}_1) \cdot (\vec{e}_1 + \vec{e}_2) = kx_2$$

Summary. The material, or Lagrangian, reference frame is a way of looking at motion where the observer follows individual material (e.g., fluid) particles as they move through space and time. Plotting the position of an individual particle through time gives the pathline of the particle. An example of this approach is if one is on a boat drifting down a river.

The spatial, or Eulerian reference, frame is a way of looking at motion that focuses on specific points in the space through which the material moves. As an illustration, consider sitting on the bank of a river and watching the water pass by your location.

Values about the motion (e.g., fluid flow) are determined as vectors at discrete locations. They are related by the material derivative

$$\frac{D\theta}{Dt} = \frac{\partial \theta}{\partial t} + \vec{v} \cdot \nabla \theta$$

which describes the rate of change of θ while moving with the material (e.g., fluid) at velocity $\vec{v}$.

2.4 STRAIN

Consider a body of particular configuration at some reference time t_0, changing to another configuration at time t as shown in Figure 2.3. We are interested in how material points close to each other change their relative positions during a deformation. Consider two points on this body, P and Q, initially close to each other, i.e., $\left| d\vec{a} \right|$ small.

Point P undergoes a displacement, $\vec{u} = \vec{x} - \vec{a}$, and arrives at its new position,

$$\vec{x} = \vec{a} + \vec{u}\,(\vec{a}, t) \tag{2.6}$$

Similarly, point Q undergoes a displacement, $\vec{u}\,(\vec{a} + d\vec{a}, t)$, and arrives at its new position,

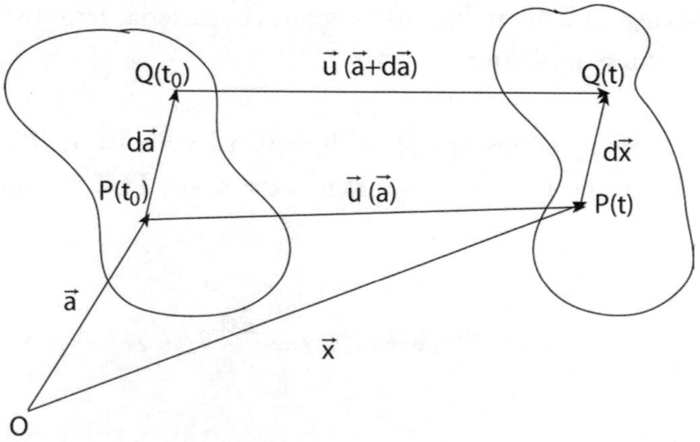

FIGURE 2.3: Deformation of a Continuum. The motion of a material particle is described by its displacement, $\vec{u}$.

$$\vec{x} + \mathrm{d}\vec{x} = \vec{a} + \mathrm{d}\vec{a} + \vec{u}(\vec{a} + \mathrm{d}\vec{a}, t) \quad (2.7)$$

Subtracting Eq. (2.6) from (2.7), we find that

$$\mathrm{d}\vec{x} = \mathrm{d}\vec{a} + \vec{u}(\vec{a} + \mathrm{d}\vec{a}, t) - \vec{u}(\vec{a}, t) \quad (2.8)$$

To understand Eq. (2.8), let us examine the first of the three equations it represents,

$$\mathrm{d}x_1 = \mathrm{d}a_1 + u_1(\vec{a} + \mathrm{d}\vec{a}, t) - u_1(\vec{a}, t) \quad (2.9)$$

Now,

$$u_1(\vec{a} + \mathrm{d}\vec{a}, t) - u_1(\vec{a}, t) = u_1(a_1 + \mathrm{d}a_1, a_2 + \mathrm{d}a_2, a_3 + \mathrm{d}a_3, t) - u_1(a_1, a_2, a_3, t)$$

can be rewritten as

$$\begin{aligned}
&u_1(a_1 + \mathrm{d}a_1, a_2 + \mathrm{d}a_2, a_3 + \mathrm{d}a_3, t) - u_1(a_1, a_2, a_3, t) = \\
&[u_1(a_1 + \mathrm{d}a_1, a_2, a_3, t) - u_1(a_1, a_2, a_3, t)] + \\
&[u_1(a_1 + \mathrm{d}a_1, a_2 + \mathrm{d}a_2, a_3, t) - u_1(a_1 + \mathrm{d}a_1, a_2, a_3, t)] + \\
&[u_1(a_1 + \mathrm{d}a_1, a_2 + \mathrm{d}a_2, a_3 + \mathrm{d}a_3, t) - u_1(a_1 + \mathrm{d}a_1, a_2 + \mathrm{d}a_2, a_3, t)]
\end{aligned}$$

Recall from multivariable calculus that $\frac{\partial F}{\partial x} = \lim_{dx \to 0} \frac{F(x + \mathrm{d}x, y, z) - F(x, y, z)}{\mathrm{d}x}$. Then,

$$\begin{aligned}
&u_1(\vec{a} + \mathrm{d}\vec{a}, t) - u_1(\vec{a}, t) = \\
&u_1(a_1 + \mathrm{d}a_1, a_2 + \mathrm{d}a_2, a_3 + \mathrm{d}a_3, t) - u_1(a_1, a_2, a_3, t) = \\
&\frac{\partial u_1}{\partial a_1}\mathrm{d}a_1 + \frac{\partial u_1}{\partial a_2}\mathrm{d}a_2 + \frac{\partial u_1}{\partial a_3}\mathrm{d}a_3
\end{aligned}$$

which leads to

$$\mathrm{d}x_1 = \mathrm{d}a_1 + u_1(\vec{a} + \mathrm{d}\vec{a}, t) - u_1(\vec{a}, t) = \mathrm{d}\vec{a}_1 + \left(\frac{\partial u_1}{\partial a_1}\mathrm{d}a_1 + \frac{\partial u_1}{\partial a_2}\mathrm{d}a_2 + \frac{\partial u_1}{\partial a_3}\mathrm{d}a_3\right) \quad (2.10)$$

Similarly,

$$\mathrm{d}x_2 = \mathrm{d}a_2 + u_2(\vec{a} + \mathrm{d}\vec{a}, t) - u_2(\vec{a}, t) = \mathrm{d}\vec{a}_2 + \left(\frac{\partial u_2}{\partial a_1}\mathrm{d}a_1 + \frac{\partial u_2}{\partial a_2}\mathrm{d}a_2 + \frac{\partial u_2}{\partial a_3}\mathrm{d}a_3\right) \quad (2.11)$$

$$\mathrm{d}x_3 = \mathrm{d}a_3 + u_3(\vec{a} + \mathrm{d}\vec{a}, t) - u_3(\vec{a}, t) = \mathrm{d}\vec{a}_3 + \left(\frac{\partial u_3}{\partial a_1}\mathrm{d}a_1 + \frac{\partial u_3}{\partial a_2}\mathrm{d}a_2 + \frac{\partial u_3}{\partial a_3}\mathrm{d}a_3\right) \quad (2.12)$$

Eqs. (2.10), (2.11), and (2.12) can be written more compactly as

$$d\vec{x} = d\vec{a} + (\nabla\vec{u})d\vec{a} \tag{2.13}$$

where the gradient of $\vec{u}$, $\nabla\vec{u}$, is identified as a tensor describing the deformation. Note that $\nabla\vec{u}$ is a tensor, but $(\nabla\vec{u})d\vec{a}$ is a vector (Eq. (1.28)). Comparing Eqs. (2.8) and (2.13), we see that $\vec{u}(\vec{a} + d\vec{a}, t) - \vec{u}(\vec{a}, t) = (\nabla\vec{u})d\vec{a}$, which is the definition of the gradient of a vector. The components of the gradient of $\vec{u}$ are given by Eq. (1.91) as

$$(\nabla\vec{u})_{ij} = \frac{\partial u_i}{\partial a_j} \tag{2.14}$$

$\nabla\vec{u}$ is the *displacement gradient*. There is a closely related tensor known as the *deformation gradient* (see Problem 3). The components of the deformation gradient tensor are given as

$$F_{ij} = \frac{\partial x_i}{\partial a_j} \tag{2.15}$$

$\underset{\sim}{F}$ is clearly not a symmetric tensor. For completeness, using $\underset{\sim}{F}$, we can define several finite deformation and finite strain tensors. These tensors are

1. The right Cauchy–Green deformation tensor, $\underset{\sim}{C}$.

$$\underset{\sim}{C} = \underset{\sim}{F}^T\underset{\sim}{F} \tag{2.16}$$

2. The left Cauchy–Green deformation tensor, $\underset{\sim}{B}$.

$$\underset{\sim}{B} = \underset{\sim}{F}\underset{\sim}{F}^T \tag{2.17}$$

3. The Eulerian strain tensor, $\underset{\sim}{\Xi}$.

$$\underset{\sim}{\Xi} = \frac{1}{2}(\underset{\sim}{I} - \underset{\sim}{B}^{-1}) \tag{2.18}$$

We will return to these equations when we discuss general elasticity. Now, consider two material vectors issuing from P, $d\vec{a}^1$ and $d\vec{a}^2$, which become through the motion, $d\vec{x}^1$ and $d\vec{x}^2$ (see Figure 2.4). Then, by Eq. (2.13),

$$\begin{aligned} d\vec{x}^1 &= d\vec{a}^1 + (\nabla\vec{u})d\vec{a}^1 \\ d\vec{x}^2 &= d\vec{a}^2 + (\nabla\vec{u})d\vec{a}^2 \end{aligned} \tag{2.19}$$

A measure of the deformation is given by the dot product of $d\vec{x}^1$ and $d\vec{x}^2$.

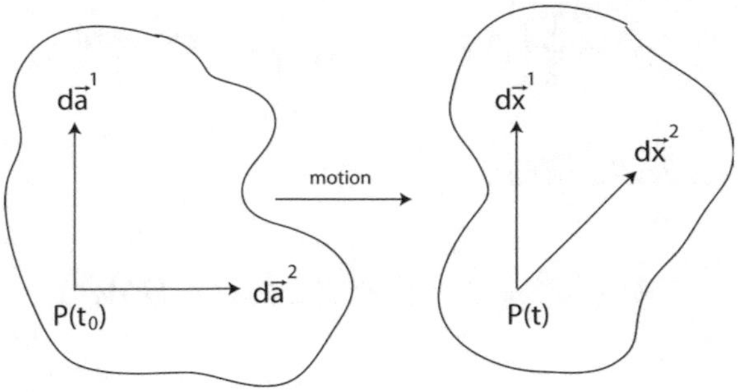

FIGURE 2.4: Relative motion of two material vectors during the deformation of a continuum.

$$d\vec{x}^1 \cdot d\vec{x}^2 = d\vec{a}^1 \cdot d\vec{a}^2 + d\vec{a}^1 \cdot (\nabla\vec{u})a\vec{a}^2 + d\vec{a}^2 \cdot (\nabla\vec{u})d\vec{a}^1 + [(\nabla\vec{u})d\vec{a}^2] \cdot [(\nabla\vec{u})d\vec{a}^1] \tag{2.20}$$

Recalling Eq. (1.48), $\vec{a}\,(\underset{\sim}{T}\vec{b}) = \vec{b}\cdot(\underset{\sim}{T}^T\vec{a})$, for the transpose of a tensor,

$$\begin{aligned} d\vec{x}^1 \cdot d\vec{x}^2 &= d\vec{a}^1 \cdot d\vec{a}^2 + d\vec{a}^1 \cdot (\nabla\vec{u})d\vec{a}^2 + d\vec{a}^1 \cdot [(\nabla\vec{u})^T d\vec{a}^2] + d\vec{a}^1 \cdot (\nabla\vec{u})^T[(\nabla\vec{u})d\vec{a}^2] \\ &= d\vec{a}^1 \cdot d\vec{a}^2 + d\vec{a}^1 \cdot \{(\nabla\vec{u}) + (\nabla\vec{u})^T + (\nabla\vec{u})^T(\nabla\vec{u})\}d\vec{a}^2 = d\vec{a}^1 \cdot d\vec{a}^2 + d\vec{a}^1 \cdot (2\underset{\sim}{E}^*)d\vec{a}^2 \end{aligned}$$

where we have defined

$$\underset{\sim}{E}^* = \frac{1}{2}\left\{(\nabla\vec{u}) + (\nabla\vec{u})^T + (\nabla\vec{u})^T(\nabla\vec{u})\right\} \tag{2.21}$$

Thus,

$$d\vec{x}^1 \cdot d\vec{x}^2 = d\vec{a}^1 \cdot d\vec{a}^2 + 2(d\vec{a}^1) \cdot (\underset{\sim}{E}^* d\vec{a}^2) \tag{2.22}$$

From Eq. (2.22), if $\underset{\sim}{E}^* = 0$, then $d\vec{x}^1 \cdot d\vec{x}^2 = d\vec{a}^1 \cdot d\vec{a}^2$, i.e., lengths of and angles between material lines remain unchanged.

$\underset{\sim}{E}^*$ is the *finite Lagrangian strain tensor*, which characterizes deformation in the neighborhood of particle P based on a material description. The components of $\underset{\sim}{E}^*$ with respect to Cartesian coordinates are

$$E^*_{ij} = \frac{1}{2}\left(\frac{\partial u_i}{\partial a_j} + \frac{\partial u_j}{\partial a_i} + \frac{\partial u_k}{\partial a_i}\frac{\partial u_k}{\partial a_j}\right) \tag{2.23}$$

For infinitesimal strain, displacements are small. Thus, the higher order product $(\nabla\vec{u})^T(\nabla\vec{u})$ can be neglected and

$$\underset{\sim}{E}^* \approx \frac{1}{2}\left\{(\nabla\vec{u}) + (\nabla\vec{u})^T\right\} = (\nabla\vec{u})^{\text{symm}} \tag{2.24}$$

So, for small deformations, Eq. (2.22) is

$$\mathrm{d}\vec{x}^1 \cdot \mathrm{d}\vec{x}^2 = \mathrm{d}\vec{a}^1 \cdot \mathrm{d}\vec{a}^2 + 2(\mathrm{d}\vec{a}^1) \cdot (\underset{\sim}{E}\,\mathrm{d}\vec{a}^2) \tag{2.25}$$

where

$$\underset{\sim}{E} \equiv \frac{1}{2}\left\{(\nabla\vec{u}) + (\nabla\vec{u})^T\right\} = (\nabla\vec{u})^{\text{symm}} \tag{2.26}$$

$\underset{\sim}{E}$ is the *infinitesimal strain tensor* (note the absence of the *). Its components, referred to a Cartesian system, are

$$E_{ij} = \frac{1}{2}\left(\frac{\partial u_i}{\partial a_j} + \frac{\partial u_j}{\partial a_i}\right) \cong \frac{1}{2}\left(\frac{\partial u_i}{\partial x_j} + \frac{\partial u_j}{\partial x_i}\right) \tag{2.27}$$

where the last equality results from $a_i \approx x_i$, as material and spatial coordinates can be assumed identical because particles do not move much during small deformations. Thus, $\underset{\sim}{E}$ is linear in the displacement gradients. Eq. (2.26) in matrix form is

$$[\underset{\sim}{E}] = \begin{bmatrix} \frac{\partial u_1}{\partial x_1} & \frac{1}{2}\left(\frac{\partial u_1}{\partial x_2} + \frac{\partial u_2}{\partial x_1}\right) & \frac{1}{2}\left(\frac{\partial u_1}{\partial x_3} + \frac{\partial u_3}{\partial x_1}\right) \\ \frac{1}{2}\left(\frac{\partial u_1}{\partial x_2} + \frac{\partial u_2}{\partial x_1}\right) & \frac{\partial u_2}{\partial x_2} & \frac{1}{2}\left(\frac{\partial u_2}{\partial x_3} + \frac{\partial u_3}{\partial x_2}\right) \\ \frac{1}{2}\left(\frac{\partial u_1}{\partial x_3} + \frac{\partial u_3}{\partial x_1}\right) & \frac{1}{2}\left(\frac{\partial u_2}{\partial x_3} + \frac{\partial u_3}{\partial x_2}\right) & \frac{\partial u_3}{\partial x_3} \end{bmatrix} \tag{2.28}$$

The diagonal elements of $\underset{\sim}{E}$ give unit elongations, i.e., *normal strains*, in the x_1, x_2, and x_3 directions. The off-diagonal elements of $\underset{\sim}{E}$ give the decrease in angle between elements, i.e., *shear strain* (e.g., $2E_{12}$ gives the decrease in angle between two elements initially in the x_1 and x_2 directions). If $\nabla\vec{u}$ is antisymmetric, recall Eq. (1.62), then $\underset{\sim}{E} = 0$, and we have an infinitesimal rigid body rotation, $\underset{\sim}{\Omega}$. The *infinitesimal rotation tensor* is given by $\underset{\sim}{\Omega} = \frac{1}{2}\left\{(\nabla\vec{u}) - (\nabla\vec{u})^T\right\}$.

Note: Under the assumption of infinitesimal deformation, the Eulerian strain tensor, Eq. (2.18), also reduces to Eq. (2.26).

2.5 PRINCIPAL STRAIN

Because the strain tensor, $\underset{\sim}{E}$, is symmetric, there exist at least three mutually perpendicular directions, $\vec{n}_1, \vec{n}_2, \vec{n}_3$ with respect to which the matrix of $\underset{\sim}{E}$ is diagonal (recall Eq. (1.76)).

$$[\underset{\sim}{E}]_{\vec{n}_1,\vec{n}_2,\vec{n}_3} = \begin{bmatrix} E_1 & 0 & 0 \\ 0 & E_2 & 0 \\ 0 & 0 & E_3 \end{bmatrix} \tag{2.29}$$

E_1, E_2, and E_3 are the *principal strains* of $\underset{\sim}{E}$. These are the eigenvalues of $\underset{\sim}{E}$, and they include the maximum and minimum normal strains. They are obtained from the characteristic equation of $\underset{\sim}{E}$, recall Eq. (1.77).

2.6 DILATATION

It can be shown (see problem 11) that

$$e = E_{ii} = \frac{\Delta(\mathrm{d}V)}{\mathrm{d}V} \tag{2.30}$$

where dV represents a small material volume. Thus, e is equal to the change in volume per unit volume. Furthermore, e is the first invariant of $\underset{\sim}{E}$, i.e., $e = E_{ii}$.

This concludes our discussion of deformation and displacement. Table 2.1 represents idealized states of deformation commonly discussed in problems or used as testing regimens.

TABLE 2.1: Simple deformations

TYPE OF DEFORMATION	x_1 or r	x_2 or θ	x_3 or z
Uniform extension or compression	$\lambda_1 a_1$	$\lambda_2 a_2$	$\lambda_3 a_3$
Uniform dilatation	$\lambda_1 a_1$	$\lambda_1 a_2$	$\lambda_1 a_3$
Simple shear	$a_1 + a_2 \tan\gamma$	a_2	a_3
Pure torsion	R	$\Theta + cZ$	Z
Pure bending	$r(a_1)$	$\theta(a_2)$	Z
Plane strain	$x_1(a_1,a_2)$	$x_2(a_1,a_2)$	a_3

In Table 2.1, λ_i's are constants known as stretch ratios; e.g., $x_1 = \lambda_1 a_1$ indicates that the body is stretched (or compressed) λ_1 times its original length in the x_1 direction. For simple shear, comparison with the demonstration on simple shearing motion should convince you that $kt = \tan\gamma$. Finally, $r(R)$, $\theta(\Theta)$, and $z(Z)$ refer to cylindrical coordinates, where $r = \sqrt{(x_1^2 + x_2^2)}$, $\theta = \tan\left(\frac{x_2}{x_1}\right)$, and $z = x_3$. In this table, pure torsion and bending are given in terms of cylindrical coordinates.

Table 2.1 gives idealized deformations without reference to a particular material model, i.e., constitutive equation. With this table, $\underset{\sim}{F}$ can be determined for a deformation, along with the several deformation and strain tensors presented. Principal strains and their corresponding directions can then be found. To determine the behavior of a material in response to any of these deformations, e.g., the stress generated in the material, we must have a constitutive equation describing the relationship between stress and strain for that particular material. We will see this in action several times later in this text.

2.7 RATE OF DEFORMATION

Consider a material element, $\mathrm{d}\vec{x}$, issuing from material point, P, located at $\vec{x}$ at time t. We wish to compute the rate of change of its length and direction. Thus, taking the material derivative of Eq. (2.8),

$$\frac{\mathrm{D}}{\mathrm{D}t}\left(\mathrm{d}\vec{x}\right) = \frac{\mathrm{D}}{\mathrm{D}t}\left(\mathrm{d}\vec{a}\right) + \frac{\mathrm{D}}{\mathrm{D}t}\left[\vec{u}(\vec{a} + \mathrm{d}\vec{a}, t) - \vec{u}(\vec{a}, t)\right] \tag{2.31}$$

Recall that the material derivative is taken with respect to a_i fixed, Eq. (2.3). Also, from basic physics, we know that the time rate of change of the displacement of a particle is its velocity. Thus, Eq. (2.31) becomes

$$\frac{\mathrm{D}}{\mathrm{D}t}\left(\mathrm{d}\vec{x}\right) = \frac{\partial}{\partial t}\left[\vec{u}\left(\vec{a} + \mathrm{d}\vec{a}, t\right) - \vec{u}\left(\vec{a}, t\right)\right] = \vec{v}\left(\vec{a} + \mathrm{d}\vec{a}, t\right) - \vec{v}\left(\vec{a}, t\right) \tag{2.32}$$

So, we get

$$\frac{\mathrm{D}}{\mathrm{D}t}\left(\mathrm{d}\vec{x}\right) = \vec{v}\left(\vec{a} + \mathrm{d}\vec{a}, t\right) - \vec{v}\left(\vec{a}, t\right) = (\nabla\vec{v})\mathrm{d}\vec{a} \tag{2.33}$$

Now, the velocity of a particle should be the same whether a material or spatial description is used, though the functional form of the velocity equation may be different. Hence, there is a velocity function $\vec{v}(\vec{x}, t)$ for which Eq. (2.33) is true when $\vec{a}$ is replaced by $\vec{x}$ on the right-hand side. Therefore, the Cartesian matrix representation of the *velocity gradient tensor*, $\nabla\vec{v}$, can be written as

$$[\nabla \vec{v}] = \begin{bmatrix} \frac{\partial v_1}{\partial x_1} & \frac{\partial v_1}{\partial x_2} & \frac{\partial v_1}{\partial x_3} \\ \frac{\partial v_2}{\partial x_1} & \frac{\partial v_2}{\partial x_2} & \frac{\partial v_2}{\partial x_3} \\ \frac{\partial v_3}{\partial x_1} & \frac{\partial v_3}{\partial x_2} & \frac{\partial v_3}{\partial x_3} \end{bmatrix} \tag{2.34}$$

$\nabla\vec{v}$ can be decomposed into the sum of its symmetric and antisymmetric parts,

$$\nabla \vec{v} = \underset{\sim}{D} + \underset{\sim}{W} \tag{2.35}$$

with

$$\underset{\sim}{D} = \frac{1}{2}\left\{(\nabla\vec{v}) + (\nabla\vec{v})^T\right\}$$

and

$$\underset{\sim}{W} = \frac{1}{2}\left\{(\nabla\vec{v}) - (\nabla\vec{v})^T\right\} \tag{2.36}$$

$\underset{\sim}{D} = (\nabla\vec{v})^{\text{symm}}$ is the rate of deformation tensor, or *strain-rate tensor*, and $\underset{\sim}{W} = (\nabla\vec{v})^{\text{asymm}}$ is the *spin tensor*. In matrix form, $\underset{\sim}{D}$ is

$$[\underset{\sim}{D}] = \begin{bmatrix} \frac{\partial v_1}{\partial x_1} & \frac{1}{2}\left(\frac{\partial v_1}{\partial x_2} + \frac{\partial v_2}{\partial x_1}\right) & \frac{1}{2}\left(\frac{\partial v_1}{\partial x_3} + \frac{\partial v_3}{\partial x_1}\right) \\ \frac{1}{2}\left(\frac{\partial v_1}{\partial x_2} + \frac{\partial v_2}{\partial x_1}\right) & \frac{\partial v_2}{\partial x_2} & \frac{1}{2}\left(\frac{\partial v_2}{\partial x_3} + \frac{\partial v_3}{\partial x_2}\right) \\ \frac{1}{2}\left(\frac{\partial v_1}{\partial x_3} + \frac{\partial v_3}{\partial x_1}\right) & \frac{1}{2}\left(\frac{\partial v_2}{\partial x_3} + \frac{\partial v_3}{\partial x_2}\right) & \frac{\partial v_3}{\partial x_3} \end{bmatrix} \tag{2.37}$$

Similar to $\underset{\sim}{E}$, the diagonal elements of $\underset{\sim}{D}$ give the rates of elongation, i.e., extension rates in the x_1, x_2, and x_3 directions. The off-diagonal elements of $\underset{\sim}{D}$ give the rate of decrease in angle between elements, i.e., shear rate (e.g., $2D_{12}$ gives the decrease in angle between two elements initially in the x_1 and x_2 directions). However, unlike for $\underset{\sim}{E}$, there are *no* approximations made in arriving at Eq. (2.37), and, therefore, Eq. (2.37) is an exact measure of the deformation rate. D_{ij} are linear in the velocity field, which simplifies the solution of many problems in fluid mechanics. We will study fluids in some detail in Chapter 5.

2.8 CONTINUITY EQUATION (CONSERVATION OF MASS)

If we follow a particle through its motion, its volume may change, but its total mass will remain unchanged. Let ρ denote the density of a particle and $\mathrm{d}V$ its volume. Then, the continuity equation says,

$$\frac{\mathrm{D}}{\mathrm{D}t}(\rho\, \mathrm{d}V) = 0 \tag{2.38}$$

Using the product rule,

$$\rho \frac{\mathrm{D}}{\mathrm{D}t}(\mathrm{d}V) + \mathrm{d}V \frac{\mathrm{D}\rho}{\mathrm{D}t} = 0$$

It can be shown that $\frac{1}{\mathrm{d}V}\frac{\mathrm{D}}{\mathrm{D}t}(\mathrm{d}V) = \frac{\partial v_i}{\partial x_i}$, the first scalar invariant of the rate of deformation tensor, $\underset{\sim}{D}$, which is the rate of change of volume per unit volume. Thus,

$$\rho \frac{\partial v_i}{\partial x_i} + \frac{\mathrm{D}\rho}{\mathrm{D}t} = 0 \tag{2.39}$$

In invariant form,

$$\rho \left(\operatorname{div} \vec{v}\right) + \frac{\mathrm{D}\rho}{\mathrm{D}t} = 0 \tag{2.40}$$

where

$$\frac{\mathrm{D}\rho}{\mathrm{D}t} = \frac{\partial \rho}{\partial t} + \vec{v} \cdot \nabla \rho$$

Demonstration. Consider the unit square OABC shown in Figure 2.5. Given the following displacement components, $u_1 = ka_2^2$ and $u_2 = u_3 = 0$,

1. Sketch the deformed shape of the unit square *OABC.*
2. Find the deformed vectors (i.e., $\mathrm{d}\vec{x}_1$ and $\mathrm{d}\vec{x}_2$) of the material elements $\mathrm{d}\vec{a}^{\,1} = \mathrm{d}a_1\, \vec{e}_1$ and $\mathrm{d}\vec{a}^{\,2} = \mathrm{d}a_2\, \vec{e}_2$ initially at point *C*.
3. Determine the ratio of the deformed to undeformed lengths of the differential elements ("stretch") of part 2 and the change in angle between these elements.
4. Let $k = 1 \times 10^{-4}$ and obtain $\underset{\sim}{E}$, the infinitesimal strain tensor.
5. Using $\underset{\sim}{E}$, find the unit elongation of material elements $\mathrm{d}\vec{a}^{\,1}$ and $\mathrm{d}\vec{a}^{\,2}$, and find the decrease in angle between these two elements.

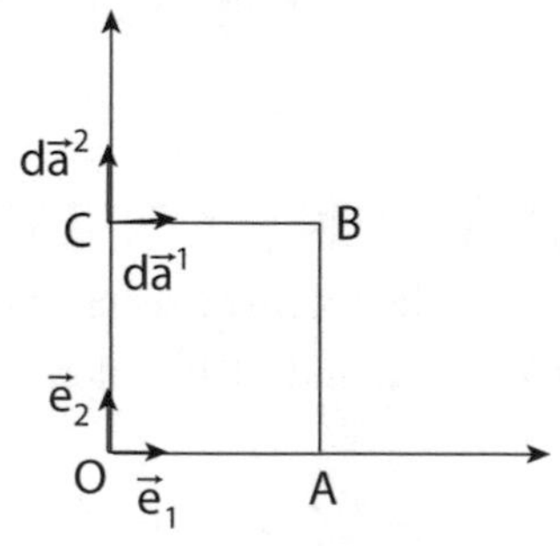

FIGURE 2.5: Undeformed unit square in the $x_1 - x_2$ plane.

Solution.

1. As in the demonstration of Section 2.2,

   Material line *OA*: $a_2 = 0 \Rightarrow u_1 = u_2 = u_3 = 0$. So, *OA* is not displaced.
   Material line *CB*: $a_2 = 1 \Rightarrow u_1 = k, = u_2 = u_3 = 0$. So, *CB* is displaced by k to the right.
   Material lines *OC* and *AB*: $u_1 = ka_2^2$. So, lines become parabolic in shape (see Figure 2.6).

2. The displacement gradient, $\nabla\vec{u}$, at point $C = (0,1,0)$ is

$$\left[\nabla\vec{u}\right]|_{@C} = \left.\begin{bmatrix} \frac{\partial u_1}{\partial a_1} & \frac{\partial u_1}{\partial a_2} & \frac{\partial u_1}{\partial a_3} \\ \frac{\partial u_2}{\partial a_1} & \frac{\partial u_2}{\partial a_2} & \frac{\partial u_2}{\partial a_3} \\ \frac{\partial u_3}{\partial a_1} & \frac{\partial u_3}{\partial a_2} & \frac{\partial u_3}{\partial a_3} \end{bmatrix}\right|_{@C} = \left.\begin{bmatrix} 0 & 2ka_2 & 0 \\ 0 & 0 & 0 \\ 0 & 0 & 0 \end{bmatrix}\right|_{a_2=1} = \begin{bmatrix} 0 & 2k & 0 \\ 0 & 0 & 0 \\ 0 & 0 & 0 \end{bmatrix}$$

Using Eq. (2.19), $\begin{matrix} \mathrm{d}\vec{x}^1 = \mathrm{d}\vec{a}^1 + \nabla\vec{u}\,\mathrm{d}\vec{a}^1 \\ \mathrm{d}\vec{x}^2 = \mathrm{d}\vec{a}^2 + \nabla\vec{u}\,\mathrm{d}\vec{a}^2 \end{matrix}$. Note: $\mathrm{d}\vec{a}^1 = \begin{bmatrix} \mathrm{d}a_1 \\ 0 \\ 0 \end{bmatrix}$ and $\mathrm{d}\vec{a}^2 = \begin{bmatrix} 0 \\ \mathrm{d}a_2 \\ 0 \end{bmatrix}$.

Therefore, $\begin{matrix} \mathrm{d}\vec{x}^1 = \mathrm{d}a_1\,\vec{e}_1 + 0 \\ \mathrm{d}\vec{x}^2 = 2k\mathrm{d}a_2\,\vec{e}_1 + \mathrm{d}a_2\,\vec{e}_2 \end{matrix}$.

3. Using Eq. (1.25), $\dfrac{\left|\mathrm{d}\vec{x}^1\right|}{\left|\mathrm{d}\vec{a}^1\right|} = \dfrac{\mathrm{d}a_1}{\mathrm{d}a_1} = 1$ and

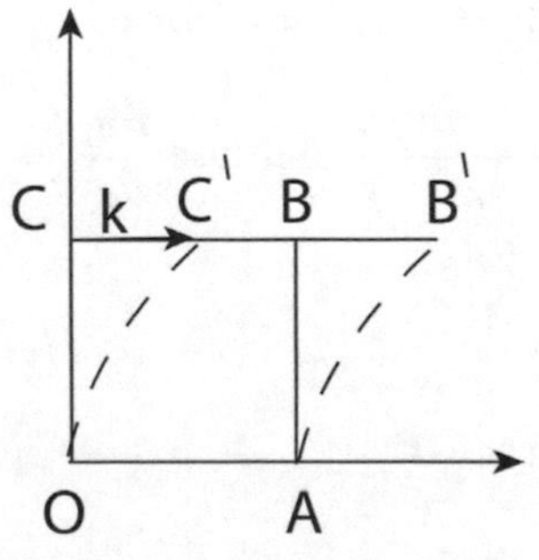

FIGURE 2.6: Deformed shape of unit square when subjected to displacement $\vec{u} = ka_2^2\,\vec{e}_1$.

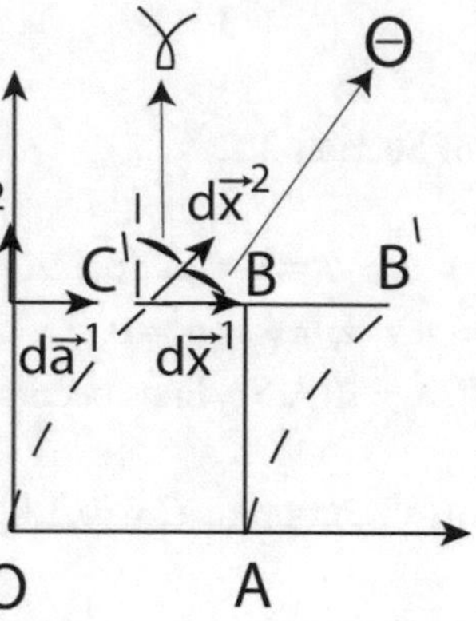

FIGURE 2.7: Definition of angles used to calculate the change in angle between $d\vec{a}^1$ and $d\vec{a}^2$. Note how $d\vec{a}^2$ rotated clockwise during the motion.

$\dfrac{\left|d\vec{x}^2\right|}{\left|d\vec{a}^2\right|} = \dfrac{\sqrt{(2k\,da_2)^2 + (da_2)^2}}{da_2} = \sqrt{4k^2+1} = 1.00000002$. Essentially, there is very little stretch.

To find the angle change, γ, recall the alternative definition of the dot product. Now, $\gamma = \dfrac{\pi}{2} - \theta$ (see Figure 2.7) and $\cos(\theta) = \dfrac{d\vec{x}^1 \cdot d\vec{x}^2}{\left|d\vec{x}^1\right|\left|d\vec{x}^2\right|}$. So,

$$\cos(\theta) = \frac{d\vec{x}^1 \cdot d\vec{x}^2}{\left|d\vec{x}^1\right|\left|d\vec{x}^2\right|} = \frac{2k\,da_1 da_2}{da_1 da_2\sqrt{4k^2+1}} = \frac{2k}{\sqrt{4k^2+1}} = \cos\left(\frac{\pi}{2} - \gamma\right) = \sin(\gamma)$$

$\gamma = \sin^{-1}\left(\dfrac{2k}{\sqrt{4k^2+1}}\right) \approx 1.9999999 \times 10^{-4}$. Note that this has been determined in radians.

4. From Eq. (2.27), $E_{ij} = \dfrac{1}{2}\left(\dfrac{\partial u_i}{\partial a_j} + \dfrac{\partial u_j}{\partial a_i}\right)$. Because $\dfrac{\partial u_1}{\partial a_2} = 2k a_2$, and all other $\dfrac{\partial u_i}{\partial a_j} = 0$,

$$[\underset{\sim}{E}] = \begin{bmatrix} 0 & k a_2 & 0 \\ k a_2 & 0 & 0 \\ 0 & 0 & 0 \end{bmatrix}, \text{ with } k = 1 \times 10^{-4}$$

5. At point C, $a_2 = 1$. So, $[\underset{\sim}{E}]\Big|_{@\ C} = \begin{bmatrix} 0 & k & 0 \\ k & 0 & 0 \\ 0 & 0 & 0 \end{bmatrix}$

Since $E_{11} = 0$ and $E_{22} = 0$, there is no elongation of material elements $d\vec{a}^1$ and $d\vec{a}^2$. The angle decrease is $2E_{12} = 2k = 2 \times 10^{-4}$. This angle decrease agrees with that calculated in part 3 as follows:

$$\gamma = \sin^{-1}\left(\frac{2k}{\sqrt{4k^2+1}}\right) \approx \sin^{-1}\left(\frac{2k}{1}\right) \approx 2k$$

where the first approximation follows from small k and the second from $\sin^{-1}(x) \approx x$ for $x \ll 1$. It should be emphasized, that if the value of k was "large" in this problem, using infinitesimal strain calculations would lead to gross error (see problem 17).

The problem solved above demonstrates two ways of quantifying local deformations around a point in a continuum. First, we can use basic principles. Based on the displacement gradient alone, we can calculate changes in unit length (normal strain) as well as changes in angles (shear strain). Second, by simply applying the definition of strain (in this case infinitesimal strain) we can obtain the same information.

2.9 PROBLEMS

1. Consider the following description of motion, $\vec{x} = \vec{a} + kta_2\vec{e}_1 + kt^3a_3\vec{e}_2$. Recall that $\vec{a}$ is the initial configuration of the body of interest.
   (a) Is this a material or spatial description of motion?
   (b) Find the inverse description. (Hint: write out the components and obtain $\vec{a}\ (\vec{x})$)
   (c) Which description of motion, material or spatial, would work best for the following situations and why?
      i. Fluid is flowing in a pipe, and you would like to know the pressures and flow rates.
      ii. An automobile crashes, and you are describing the motion of the head of a crash test dummy.

2. Consider the motion $\vec{x} = \vec{a} + ka_2^3t\vec{e}_1 - 5ka_1^2t\vec{e}_2$ with a pressure profile described by $p_g = -\rho g x_2$.
   (a) Find the material description of p_g.

(b) Obtain the material description of the velocity vector.
(c) Obtain the rate of pressure change.
(d) Use Matlab© to sketch the configuration at time $t = 0$ of a rectangular prism with edges at the coordinate axes and at $a_1 = 1$, $a_2 = 2$ and its resulting deformation given $k = 1 \times 10^{-4}$ at time $t = 1{,}000$.

3. In this problem, we will examine various relationships among the deformation and displacement gradients and the Eulerian, Lagrangian, and infinitesimal strain tensors.

(a) Knowing that $\vec{u} = \vec{x} - \vec{a}$, show that $\nabla \vec{u} = \underset{\sim}{F} - \underset{\sim}{I}$, where the gradient is taken with respect to material coordinates.
(b) Use (a) to write the Lagrangian strain tensor, $\underset{\sim}{E}^*$, in terms of $\underset{\sim}{F}$. What is it in terms of $\underset{\sim}{C}$?
(c) Write the infinitesimal strain tensor, $\underset{\sim}{E}$, in terms of $\underset{\sim}{F}$. In words, how does this expression demonstrate that $\underset{\sim}{E}$ is symmetric?
(d) Write the Eulerian strain tensor, $\underset{\sim}{\Xi}$, in terms if $\underset{\sim}{F}$.
(e) Finally, show $\underset{\sim}{E}^* = \underset{\sim}{F}^T \underset{\sim}{\Xi} \underset{\sim}{F}$.

Hint: The following two expressions will be useful for solving the above problem.
Given two tensors, $\underset{\sim}{S}$ and $\underset{\sim}{T}$, $(\underset{\sim}{S} + \underset{\sim}{T})^T = \underset{\sim}{S}^T + \underset{\sim}{T}^T$ and $(\underset{\sim}{S}\underset{\sim}{T})^{-1} = \underset{\sim}{T}^{-1}\underset{\sim}{S}^{-1}$.

4. The components of $\underset{\sim}{F}^{-1}$ are $\left(F^{-1}\right)_{ij} = \frac{\partial a_i}{\partial x_j}$. The Lagrangian strain tensor is a function of $\underset{\sim}{C}$, while the Eulerian strain tensor is a function of $\underset{\sim}{B}^{-1}$. Based on what you know about the components of $\underset{\sim}{F}$ and $\underset{\sim}{F}^{-1}$, explain why the names of these tensors make sense.

5. Prove the following relationships between the invariants of $\underset{\sim}{C}$ and $\underset{\sim}{E}^*$. Chapter 1 problems 17 and 18 may help.
(a) $I_1^{C} = 3 + 2I_1^{E^*}$
(b) $I_2^{C} = 3 + 4I_1^{E^*} + 4I_2^{E^*}$
(c) $I_3^{C} = 1 + 2I_1^{E^*} + 4I_2^{E^*} + 8I_3^{E^*}$

Note: The tensor to which the invariant refers has been superscripted.

6. Take the case of simple shear from Table 2.1.

(a) Calculate $\underset{\sim}{E}^*$. (Hint: Start by calculating $\underset{\sim}{F}$).
(b) Calculate $\underset{\sim}{E}$. Comment on why this is different from $\underset{\sim}{E}^*$.
(c) Show that $\underset{\sim}{E}^*$ reduces to $\underset{\sim}{E}$ for small deformation (i.e., γ small). You will need the fact that the Taylor expansion of $\tan\gamma = \gamma + \frac{\gamma^3}{3} + \frac{2\gamma^5}{15} + \ldots \quad \text{for} \quad |\gamma| < \frac{\pi}{2}$.
(d) Calculate the principal strains and principal directions for $\underset{\sim}{E}$ given a simple shear deformation.

7. The velocity gradient tensor can be decomposed into its symmetric and antisymmetric parts, the rate of deformation and spin tensors, respectively. In this problem, you will investigate the decomposition of the displacement gradient tensor. Given a displacement field $\vec{u}$,
   (a) Calculate the gradient of $\vec{u}$.
   (b) Decompose this tensor into symmetric and antisymmetric parts.
   (c) What does the symmetric part represent?

   Note: The antisymmetric part is the infinitesimal rotation tensor.

8. This problem looks at the *infinitesimal rotation vector*. The infinitesimal rotation vector is defined as

$$\vec{\omega} = \frac{1}{2}\text{curl}\left(\vec{u}\right) \tag{2.41}$$

   Recall that curl $(\vec{u}) \equiv \nabla \times \vec{u}$.
   (a) Show that $\vec{\omega} = -\frac{1}{2}\varepsilon_{ijk}\Omega_{jk}\vec{e}_i$, where $\underset{\sim}{\Omega}$ is the infinitesimal rotation tensor.
   (b) Based on the above result, given a general vector $\vec{v}$, what is the formula for the curl of $\vec{v}$ in terms of $(\nabla\vec{v})^{\text{asymm}}$?

9. The continuity equation can be derived from a simple mass balance (i.e., rate of accumulation = rate in − rate out) in the form $\frac{\partial\rho}{\partial t} + \nabla\cdot(\rho\vec{v}) = 0$. Starting from this equation, derive the material derivative form of the continuity equation, $\frac{\mathrm{D}\rho}{\mathrm{D}t} + \rho\frac{\partial v_i}{\partial x_i} = 0$. Write this equation out in its full form, i.e., expand the material derivative and write out the dot product and the summation over i.

10. It can be shown that the relationship between the initial ($\mathrm{d}V_0$) and deformed ($\mathrm{d}V$) volumes is

$$dV = \left(\det \underset{\sim}{F}\right) dV_0 \tag{2.42}$$

(a) Based on the principle of conservation of mass, show that Eq. (2.42) implies $\rho = \frac{\rho_0}{\det \underset{\sim}{F}}$.

(b) For an incompressible material (i.e., volume remains constant throughout the deformation), what is $\det \underset{\sim}{F}$? What is the relationship among the principal values of $\underset{\sim}{F}$ in this case?

Note: $\det \underset{\sim}{F}$ is commonly referred to simply as J.

(c) Given a uniform dilatation with $\lambda = ½$ (i.e., a compression in this case) and an initial volume of 2 cm^3, what is the volume after the deformation?

11. Show that the dilatation is given by, $e = E_{ii} = \frac{\Delta(dV)}{dV}$. *In words*, what does dilatation measure?

12. Consider the velocity field $\vec{v} = \left(\frac{x_1}{1+t}\right)\vec{e}_1$.
    (a) An incompressible material has constant density, i.e., ρ is independent of space and time. What is the continuity equation for an incompressible material? Can the above velocity field belong to an incompressible material?
    (b) Now consider a material where the density is independent of spatial position but not independent of time, i.e., $\rho = \rho(t)$. For the above velocity field, what is the density as a function of time?

13. A rod with circular cross-section is twisted such that the following displacement field is developed: $u_1 = -\psi x_2 x_3 \quad u_2 = \psi x_1 x_3 \quad u_3 = 0$, with ψ constant. Calculate the infinitesimal strain and infinitesimal rotation tensors. Find the principal strains.

14. A unit cube, with edges parallel to the coordinate axes, is subjected to the following displacement field: $u_1 = -ka_1^2 \quad u_2 = k^{1/2} a_2 \quad u_3 = 0$, with $k = 1 \times 10^{-5}$.
    (a) Find the increase in length of the diagonal AB, where A is originally at (0, 0, 0) and B is originally at (1, 1, 0).
    (b) What is the percent change in volume of the cube?
    (c) Sketch the result.

15. Consider the displacement field $u_1 = k^{1/2} a_1 a_3$, $u_2 = k(a_2 + 2a_1)$, and $u_3 = ka_1^2$, with $k = 1 \times 10^{-5}$. For an element at the point (1, 1, 2) with sides parallel to the coordinate axes,

(a) What is the unit elongation in the x_2 direction? *In words*, what does elongation measure?

(b) Find the maximum unit elongation for this element. How would you find the direction associated with this elongation?

(c) What is the change in angle between the sides initially parallel to the x_1 and x_3 axes, respectively?

16. A unit cube, with edges parallel to the coordinate axes, is subject to the displacement field $u_1 = k^2(2a_1 + a_2^2), u_2 = k(a_1^2 - a_2^2)$, and $u_3 = 0$.

(a) Given the material vector $\mathrm{d}\vec{a} = \mathrm{d}a_1\vec{e}_1 + \mathrm{d}a_2\vec{e}_2$ originating from the point (1,1,0), find the ratio $\dfrac{|\mathrm{d}\vec{x}|}{|\mathrm{d}\vec{a}|}$. You do not need to simplify these expressions.

(b) Find the infinitesimal strain tensor.

(c) At the point (1,1,0), what is the change in volume if $k = 1 \times 10^{-3}$?

(d) What is the change in angle between material elements initially parallel to the x_2 and x_3 axes?

17. In the last demonstration of this chapter, we let $k = 1 \times 10^{-4}$. Recalculate the strain and change in angle for $k = 1 \times 10^{-2}$, $k = 1$, and $k = 10$ using both basic principles and infinitesimal strain. What is the percent error in strain and angle change when using the infinitesimal strain assumption for each of these values of k? Comment on the differences. (Hint: What is the relationship between stretch and strain?)

• • • •

CHAPTER 3

Stress

Internal forces in real matter are those among molecules, atoms, and subatomic particles. In continuum theory, internal forces are considered to be

1. *Body forces*—These are forces that act throughout the volume of a body (e.g., gravity, electrostatic).
2. *Surface forces*—These forces act on a surface (real or imagined), separating parts of the body (e.g., due to contact with another body, i.e., pressure).

3.1 STRESS VECTOR ("TRACTION")

The surface force at a point on a surface can be described by the stress vector. Consider a body as shown in Figure 3.1.

A plane, S, with normal unit vector $\vec{n}$ passes through an arbitrary point, P. Consider a portion of the original body to be a free body. Then, $\Delta\vec{F}$ is the resultant force acting on the small area ΔA around P. Let us define the stress vector, $\vec{t}_{\vec{n}}$, at point P on plane S as

$$\vec{t}_{\vec{n}} = \lim_{\Delta A \to 0} \frac{\Delta\vec{F}}{\Delta A} \tag{3.1}$$

$\vec{t}_{\vec{n}}$ has units of force over area (e.g., N/mm^2 = MPa).

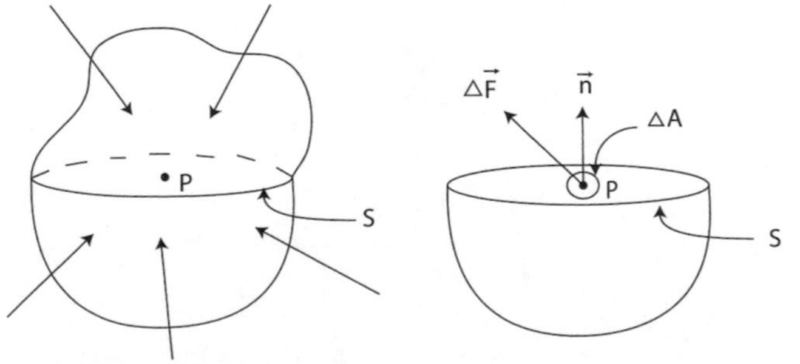

FIGURE 3.1: Pictorial depiction of the stress vector.

3.2 STRESS TENSOR AND ITS COMPONENTS

If $\vec{n}$ is a unit normal vector to a plane, then the stress vector, $\vec{t}_{\vec{n}}$, on the plane is given by Cauchy's formula

$$\vec{t}_{\vec{n}} = \underset{\sim}{T}\vec{n} \tag{3.2}$$

where the stress tensor, $\underset{\sim}{T}$, is a linear transformation (second-order tensor) known as the Cauchy stress tensor.

From $\vec{t}_{\vec{e}_1} = \underset{\sim}{T}\vec{e}_1 = T_{11}\vec{e}_1 + T_{21}\vec{e}_2 + T_{31}\vec{e}_3$, T_{11} is the normal component (normal stress) of the stress vector $\vec{t}_{\vec{e}_1}$ on the plane whose normal is $\vec{e}_1$. T_{21} and T_{31} are the in-plane components (shear stress) of the stress vector $\vec{t}_{\vec{e}_1}$ on the plane whose normal is $\vec{e}_1$ (see Figure 3.2).

So, the diagonal elements of $\underset{\sim}{T}$, T_{11}, T_{22}, T_{33}, are the *normal stresses*. The off-diagonal elements of $\underset{\sim}{T}$, T_{12}, T_{13}, T_{21}, T_{23}, T_{31}, T_{32}, are the *shear stresses*. When looking at the components of the stress tensor, keep in mind what the subscripts mean. The first number denotes direction, while the second number denotes the unit normal of the plane on which it is acting.

3.3 PRINCIPLE OF MOMENT OF MOMENTUM (PROOF OF STRESS TENSOR SYMMETRY)

We will show that the stress tensor, $\underset{\sim}{T}$, is generally a symmetric tensor by the use of the moment of momentum equations for a differential element. Consider the free body diagram of a differential parallelepiped shown in Figure 3.3. We will find the moment of all forces about an axis passing through the center point, A, and parallel to the x_3-axis.

For equilibrium, $\sum M_A = 0$, where we take the positive moment defined as counter-clockwise with thumb pointing out of the page. Then,

$$\sum M_A = T_{21}(\Delta x_2 \Delta x_3)\left(\frac{\Delta x_1}{2}\right) + (T_{21} + \Delta T_{21})(\Delta x_2 \Delta x_3)\left(\frac{\Delta x_1}{2}\right)$$
$$-T_{12}(\Delta x_1 \Delta x_3)\left(\frac{\Delta x_2}{2}\right) - (T_{12} + \Delta T_{12})(\Delta x_1 \Delta x_3)\left(\frac{\Delta x_2}{2}\right) = 0$$

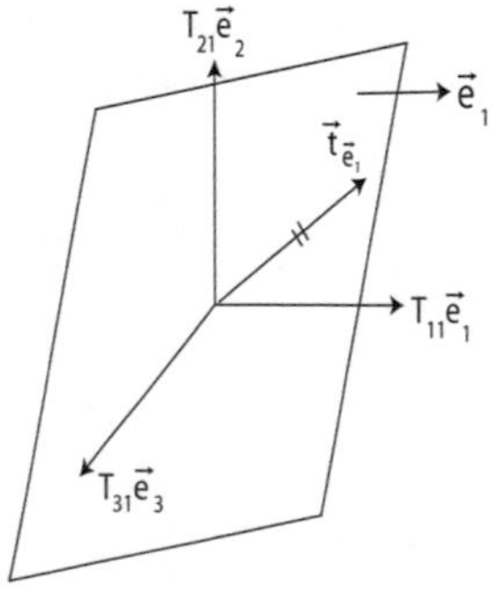

FIGURE 3.2: Components of the stress vector on a plane defined by unit normal $\vec{e}_1$.

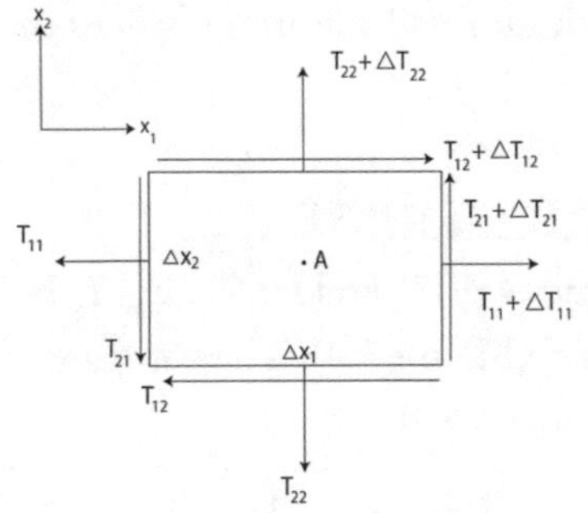

FIGURE 3.3: Balance of the moments of momentum acting on a parallelpiped about point A.

Dropping terms containing small quantities of higher order,

$$\sum M_A = (\,T_{21} - T_{12}\,)(\Delta x_1 \Delta x_2 \Delta x_3\,) = 0$$

for equilibrium. Thus,

$$T_{12} = T_{21} \tag{3.3}$$

Similarly, considering the other directions, $T_{13} = T_{31}$ and $T_{23} = T_{32}$. Hence, there are only six independent components of $\underset{\sim}{T}$. Also, $\underset{\sim}{T}$ is symmetric. Though we have shown the stress tensor is symmetric for an equilibrium situation without body forces, it is true in general (see problem 10). An exception to stress tensor symmetry is for a special class of materials known as polar materials. We will not consider such materials.

3.4 PRINCIPAL STRESSES

For any symmetric tensor, there exists at least three mutually perpendicular principal directions (eigenvectors). The planes having these directions as their normals are the *principal planes*. On these planes, the stress vector is normal to the plane (i.e., no shear stresses). The normal stresses are the *principal stresses* (eigenvalues), obtained from the characteristic equation of $\underset{\sim}{T}$ (recall Eq. (1.77)).

$$\lambda^3 - I_1\lambda^2 + I_2\lambda - I_3 = 0 \tag{3.4}$$

where the individual scalar invariants, I_i's, are given by

$$\begin{aligned} I_1 &= T_{11} + T_{22} + T_{33} \\ I_2 &= \begin{vmatrix} T_{11} & T_{12} \\ T_{21} & T_{22} \end{vmatrix} + \begin{vmatrix} T_{11} & T_{13} \\ T_{31} & T_{33} \end{vmatrix} + \begin{vmatrix} T_{22} & T_{23} \\ T_{32} & T_{33} \end{vmatrix} \\ I_3 &= \det(\underset{\sim}{T}) = (T_{11}T_{22}T_{33}) + 2(T_{12}T_{23}T_{13}) - (T_{11}T_{23}^2 + T_{22}T_{13}^2 + T_{33}T_{12}^2) \end{aligned} \tag{3.5}$$

Principal stresses include the maximum and minimum values of normal stresses among all planes passing through a given point.

3.5 MAXIMUM SHEAR STRESS

Let $\vec{e}_1, \vec{e}_2, \vec{e}_3$ be the principal directions of $\underset{\sim}{T}$, and let T_1, T_2, T_3 be the principal stresses. Consider a plane whose unit normal is $\vec{n} = n_i\vec{e}_i$ (Figure 3.4). The components of the stress vector on the plane, from Eq. (3.2), are given by (recall Eq. (1.76))

$$\begin{bmatrix} t_1 \\ t_2 \\ t_3 \end{bmatrix} = \begin{bmatrix} T_1 & 0 & 0 \\ 0 & T_2 & 0 \\ 0 & 0 & T_3 \end{bmatrix} \begin{bmatrix} n_1 \\ n_2 \\ n_3 \end{bmatrix} = \begin{bmatrix} n_1 T_1 \\ n_2 T_2 \\ n_3 T_3 \end{bmatrix} \tag{3.6}$$

or

$$\vec{t}_{\vec{n}} = n_1 T_1 \vec{e}_1 + n_2 T_2 \vec{e}_2 + n_3 T_3 \vec{e}_3 \tag{3.7}$$

Recall from vector analysis that the projection of a vector $\vec{u}$ onto a vector $\vec{v}$, $\text{proj}_{\vec{v}}\vec{u}$, is given by $\text{proj}_{\vec{v}}\,\vec{u} = \left(\dfrac{\vec{u}\cdot\vec{v}}{|\vec{v}|^2}\right)\vec{v}$ (Figure 3.5). In our case, the normal stress is given by $\text{proj}_{\vec{n}}\vec{t}_{\vec{n}}$.

$$\text{proj}_{\vec{n}}\,\vec{t}_{\vec{n}} = \left(\frac{\vec{t}_{\vec{n}}\cdot\vec{n}}{|\vec{n}|^2}\right)\vec{n} = \vec{T}_n \tag{3.8}$$

Noting that $|\vec{n}| = 1$, the magnitude of the normal stress, $|\vec{T}_n|$, on the plane defined by $\vec{n}$ is given by

$$\left|\vec{T}_n\right| = \vec{t}_{\vec{n}}\cdot\vec{n} = n_1^2 T_1 + n_2^2 T_2 + n_3^2 T_3 \tag{3.9}$$

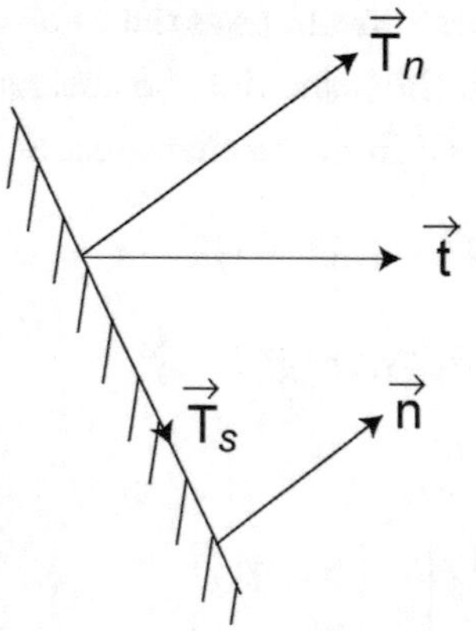

FIGURE 3.4: Decomposition of the stress vector into normal and shear components on a plane defined by unit normal $\vec{n}$.

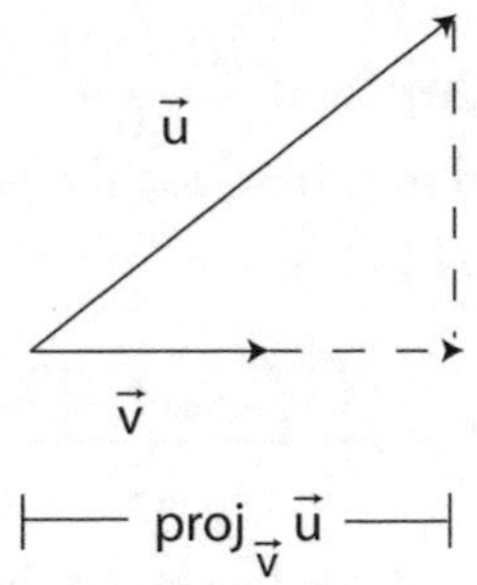

FIGURE 3.5: Projection of a vector $\vec{u}$ on to a vector $\vec{v}$.

Then, the magnitude of the total shear stress, $|\vec{T}_s|$, on the plane is (using the Pythagorean Theorem)

$$|\vec{T}_S| = \sqrt{(|\vec{t}_{\vec{n}}|^2 - |\vec{T}_n|^2)} \tag{3.10}$$

Eqs. (3.7) and (3.9) into Eq. (3.10) yields

$$|\vec{T}_S|^2 = n_1^2 T_1^2 + n_2^2 T_2^2 + n_3^2 T_3^2 - (n_1^2 T_1 + n_2^2 T_2 + n_3^2 T_3)^2 \tag{3.11}$$

To obtain the values of (n_1, n_2, n_3) for which $|\vec{T}_s|^2$ is a maximum or minimum, we find values for (n_1, n_2, n_3) such that $|\vec{T}_s|^2$ is stationary. It can be shown [2] that the values of *minimum shear stress*, specifically $|\vec{T}_s| = 0$, occur on the principal planes.

$$\vec{n} = (1,0,0),\ (0,1,0),\ \text{or}\ (0,0,1) \tag{3.12}$$

The *maximum shear stress* occurs on one of the planes given by

$$\vec{n} = \left(\frac{1}{\sqrt{2}}, \pm\frac{1}{\sqrt{2}}, 0\right), \left(\frac{1}{\sqrt{2}}, 0, \pm\frac{1}{\sqrt{2}}\right), \text{or} \left(0, \frac{1}{\sqrt{2}}, \pm\frac{1}{\sqrt{2}}\right) \tag{3.13}$$

On the planes given by Eq. (3.13), the values of $|\vec{T}_s|^2$ are, from Eq. (3.11),

$$\begin{aligned}
|\vec{T}_S|^2 &= \frac{(T_1 - T_2)^2}{4} \quad \text{for} \quad \vec{n} = \frac{1}{\sqrt{2}}\vec{e}_1 \pm \frac{1}{\sqrt{2}}\vec{e}_2 \\
|\vec{T}_S|^2 &= \frac{(T_1 - T_3)^2}{4} \quad \text{for} \quad \vec{n} = \frac{1}{\sqrt{2}}\vec{e}_1 \pm \frac{1}{\sqrt{2}}\vec{e}_3 \\
|\vec{T}_S|^2 &= \frac{(T_2 - T_3)^2}{4} \quad \text{for} \quad \vec{n} = \frac{1}{\sqrt{2}}\vec{e}_2 \pm \frac{1}{\sqrt{2}}\vec{e}_3
\end{aligned} \tag{3.14}$$

Thus, the maximum shear stress is the largest of $\frac{|T_1 - T_2|}{2}$, $\frac{|T_1 - T_3|}{2}$, or $\frac{|T_2 - T_3|}{2}$.

From this example, it can be inferred that, for the general case, the maximum shear stress corresponding to a stress tensor is given by

$$|\vec{T}_S|^{\max} = \frac{|T^{\max}_{\text{Principal}} - T^{\min}_{\text{Principal}}|}{2} \tag{3.15}$$

Further, on the plane of maximum shear, it can be shown that the normal stress is

$$|\vec{T}_n| = \frac{|T^{\max}_{\text{Principal}} + T^{\min}_{\text{Principal}}|}{2} \tag{3.16}$$

Summary.

1. The maximum shear stress is equal to one-half the difference between the maximum and minimum principal stresses.
2. It acts on the plane that bisects the angle between the directions of the maximum and minimum principal stresses.
3. The principal planes have zero shear stress.

Table 3.1 shows the components of the stress tensor for some idealized loading conditions. In this table, p, σ, τ, and c are constants and $r^2 = x_2^2 + x_3^2$. This table allows one to form the stress tensor, $\underset{\sim}{T}$, for each of these states of stress, and thus find the principal stresses, maximum and minimum

TABLE 3.1: States of stress for an object with its axis coincident with the x_1 axis

LOADING CONDITION	T_{11}	T_{12}	T_{13}	T_{22}	T_{23}	T_{33}
Hydrostatic pressure	$-p$	0	0	$-p$	0	$-p$
Uniform tension or compression	σ	0	0	0	0	0
Simple shear	0	τ	0	0	0	0
Pure bending	cx_2	0	0	0	0	0
Pure torsion	0	$T_{12} = -x_3 f(r)$	$T_{13} = x_2 f(r)$	0	0	0
Plane stress	$T_{11} = T_{11}(x_1, x_2)$	$T_{12} = T_{12}(x_1, x_2)$	0	$T_{22} = T_{22}(x_1, x_2)$	0	0

shear stresses, and normal stress on the plane of maximum shear. As it was for Table 2.1, this table does not refer to any particular material model. We need a constitutive equation to determine the strains generated in a given material in response to any of these states of stress.

3.6 EQUATIONS OF MOTION (CONSERVATION OF LINEAR MOMENTUM)

We wish to determine the differential equations of motion for any continuum in motion. Every particle of the continuum must satisfy Newton's law of motion. Consider the stress vectors acting on the six faces of a small rectangular element of the continuum as shown in Figure 3.6.

Let $\vec{B} = B_i \vec{e}_i$ be the body force (e.g., weight) per unit mass, ρ = mass density at x_i, and $\vec{a}$ = acceleration at position x_i. Applying Newton's second law of motion,

$$\begin{aligned} &\left\{ \frac{1}{\Delta x_1} \left[\vec{t}_{\vec{e}_1}(x_1 + \Delta x_1, x_2, x_3) - \vec{t}_{\vec{e}_1}(x_1, x_2, x_3) \right] \right. \\ &+ \frac{1}{\Delta x_2} \left[\vec{t}_{\vec{e}_2}(x_1, x_2 + \Delta x_2, x_3) - \vec{t}_{\vec{e}_2}(x_1, x_2, x_3) \right] \\ &\left. + \frac{1}{\Delta x_3} \left[\vec{t}_{\vec{e}_3}(x_1, x_2, x_3 + \Delta x_3) - \vec{t}_{\vec{e}_3}(x_1, x_2, x_3) \right] + \rho \vec{B} \right\} \Delta x_1 \Delta x_2 \Delta x_3 \\ &= (\rho \vec{a}) \Delta x_1 \Delta x_2 \Delta x_3 \end{aligned} \tag{3.17}$$

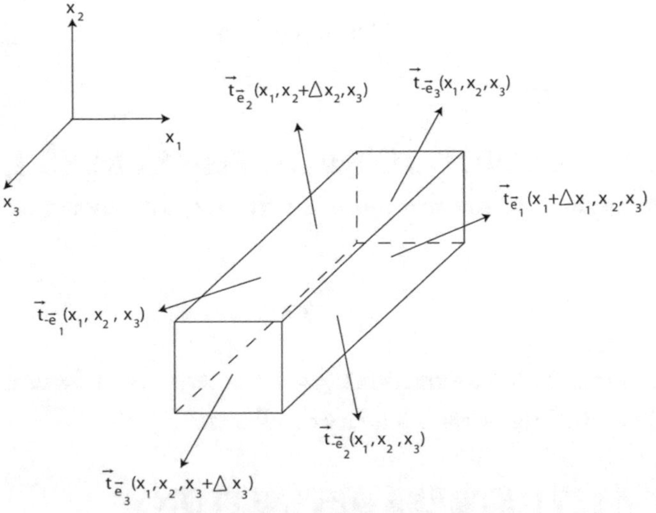

FIGURE 3.6: Stress vectors acting on the six faces of a rectangular parallelepiped located within a continuum. Note that the stress vector on each face contains both normal and shear components.

Dividing Eq. (3.17) by $\Delta x_1\,\Delta x_2\,\Delta x_3$ and taking the $\lim\limits_{\Delta x_i \to 0}$,

$$\frac{\partial \vec{t}_{\vec{e}_1}}{\partial x_1} + \frac{\partial \vec{t}_{\vec{e}_2}}{\partial x_2} + \frac{\partial \vec{t}_{\vec{e}_3}}{\partial x_3} + \rho \vec{B} = \rho \vec{a} \tag{3.18}$$

But, knowing $\vec{t}_{\vec{e}_j} = \underset{\sim}{T}\,\vec{e}_j = T_{ij}\,\vec{e}_i$, Eq. (3.18) becomes

$$\frac{\partial T_{ij}}{\partial x_j}\vec{e}_i + \rho B_i \vec{e}_i = \rho a_i \vec{e}_i \tag{3.19}$$

In invariant form,

$$\mathrm{div}(\underset{\sim}{T}) + \rho \vec{B} = \rho \vec{a} \tag{3.20}$$

which, in component form for a Cartesian coordinate system, is

$$\frac{\partial T_{ij}}{\partial x_j} + \rho B_i = \rho a_i \tag{3.21}$$

Eqs. (3.20) and (3.21) are known as Cauchy's equations of motion. If acceleration vanishes (e.g., quasi-static conditions) Eqs. (3.20) and (3.21) reduce to the equilibrium equations

$$\begin{gathered} \mathrm{div}(\underset{\sim}{T}) + \rho \vec{B} = 0 \\ \frac{\partial T_{ij}}{\partial x_j} + \rho B_i = 0 \end{gathered} \tag{3.22}$$

3.7 BOUNDARY CONDITION FOR THE STRESS TENSOR

Surface tractions are applied distributive forces on the actual boundary of some body. We can show that

$$\vec{t}_{\vec{n}} = \underset{\sim}{T}\vec{n} \tag{3.23}$$

$\vec{t}_{\vec{n}}$ = force vector per unit area (surface traction), $\underset{\sim}{T}$ = stress tensor, and $\vec{n}$ = unit vector normal to the boundary. Eq. (3.23) is called the stress boundary condition.

*3.8 ALTERNATIVE STRESS DEFINITIONS

In Section 3.2, we introduced the Cauchy stress tensor. This tensor is referred to the current configuration, x_i's. Furthermore, Cauchy's equations of motion refer to the current configuration (spatial description), such that quantities, e.g., ρ, are evaluated in the current configuration as well. Other measures of stress refer back to the body's original configuration (material description), and they can be expressed in terms of $\underset{\sim}{T}$.

Let us define the surface traction on a given surface referred to the body's original configuration as

$$\vec{t}_{\vec{n}_0} \equiv \underset{\sim}{\mathrm{P}}\vec{n}_0 \tag{3.24}$$

where $\underset{\sim}{\mathrm{P}}$ is the first Piola–Kirchoff, also known as the nominal or Lagrangian, stress tensor, and $\vec{n}_0$ is the unit normal to that surface in the reference configuration. Then, the resultant force on that surface is equal to the traction times the area of the surface. In the limit, as the area of the surface goes to zero,

$$\mathrm{d}\vec{f} = \vec{t}_{\vec{n}_0}\mathrm{d}A_0 \tag{3.25}$$

In the current configuration, the differential force acting on the same surface is

$$\mathrm{d}\vec{f} = \vec{t}_{\vec{n}}\mathrm{d}A \tag{3.26}$$

The resultant force on a surface does not depend on the description used. Because the force must be equal, the two expressions for $\mathrm{d}\vec{f}$ must also be equal. Equating Eqs. (3.25) and (3.26), and substituting the definitions for $\vec{t}_{\vec{n}}$ and $\vec{t}_{\vec{n}_0}$, we get

$$\vec{t}_{\vec{n}_0}\mathrm{d}A_0 = \vec{t}_{\vec{n}}\mathrm{d}A = \underset{\sim}{\mathrm{P}}\vec{n}_0\mathrm{d}A_0 = \underset{\sim}{T}\vec{n}\mathrm{d}A \tag{3.27}$$

It can be shown that $\vec{n}_0\mathrm{d}A_0 = \dfrac{1}{(\det\underset{\sim}{F})}\underset{\sim}{F}^T\vec{n}\mathrm{d}A$. Thus, Eq. (3.27) becomes

$$\underset{\sim}{\mathrm{P}}\frac{1}{(\det\underset{\sim}{F})}\underset{\sim}{F}^T\vec{n}\mathrm{d}A = \underset{\sim}{T}\vec{n}\mathrm{d}A \tag{3.28}$$

from which we see $\underset{\sim}{\mathrm{P}}$ is related to $\underset{\sim}{T}$ by

$$\underset{\sim}{\mathrm{P}} = (\det\underset{\sim}{F})\underset{\sim}{T}\underset{\sim}{F}^{-T} \tag{3.29}$$

where $\underset{\sim}{F}$ is the deformation gradient introduced in Eq. (2.15). The equations of motion can be expressed in terms of $\underset{\sim}{\mathrm{P}}$ as

$$\left(\frac{\partial \mathrm{P}_{ij}}{\partial a_j}\right) + \rho_0 B_i = \rho_0 \alpha_i \tag{3.30}$$

where a_j's are the material coordinates of the points in the body, ρ_0 the original density, and α the material element's acceleration (α used here to not confuse the acceleration with the material coordinates). Finally, another closely related stress tensor, the second Piola–Kirchoff stress tensor, is formed from the first Piola–Kirchoff stress tensor by

$$\underset{\sim}{P}_2 = F^{-1}\underset{\sim}{P} \tag{3.31}$$

The Cauchy stress tensor is related to the first and second Piola–Kirchoff stress tensors by

$$\underset{\sim}{T} = J^{-1}\underset{\sim}{P}\underset{\sim}{F}^T = J^{-1}\underset{\sim}{F}\underset{\sim}{P}_2\underset{\sim}{F}^T \tag{3.32}$$

where we have used the common notation $J = \det \underset{\sim}{F}$ (Chapter 2, problem 10). The second Piola–Kirchoff stress tensor is symmetric if $\underset{\sim}{T}$ is symmetric, whereas the first Piola–Kirchoff stress tensor is, in general, not. It can be more convenient to express some constitutive equations using the second Piola–Kirchoff stress tensor.

3.9 DEMONSTRATIONS

1. The state of stress at a certain point is $\underset{\sim}{T} = -p\underset{\sim}{I}$, where p is a scalar. Show that there is no shear on any plane containing this point.

Solution. From Eq. (3.2), $\vec{t}_{\vec{n}} = \underset{\sim}{T}\vec{n} \Rightarrow \vec{t}_{\vec{n}} = -p\underset{\sim}{I}\vec{n} = -p\vec{n}$. This means that the stress vector is normal to the plane, i.e., there is no shear component. This state of stress is called *hydrostatic pressure*.

2. Consider a rectangular block inside a body (see Figure 3.7).

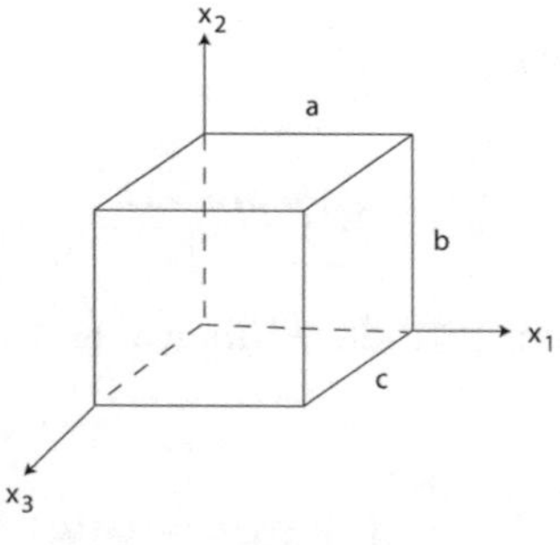

FIGURE 3.7: Rectangular block with sides parallel to the coordinate axes.

The distribution of stress inside the body is given by

$$\left[\underset{\sim}{T}\right] = \begin{bmatrix} -p+\rho g x_2 & 0 & 0 \\ 0 & -p+\rho g x_2 & 0 \\ 0 & 0 & -p+\rho g x_2 \end{bmatrix}, \text{ or } T_{ij} = (-p+\rho g x_2)\delta_{ij} \text{ with constant } g, \rho, \text{ and } p.$$

(a) What is the distribution of the stress vector on the six faces of the block (see Figure 3.8)?

(b) Find the total resultant forces acting on faces $x_2 = 0$ and $x_1 = 0$.

Solution.

(a) From Eq. (3.2), $\vec{t}_{\vec{n}} = \underset{\sim}{T}\vec{n}$

On

$x_1 = 0$	$\vec{n} = (-1,0,0)$,	$\vec{t}_{\vec{n}} = (p - \rho g x_2, 0, 0)$
$x_1 = \text{a}$	$\vec{n} = (1,0,0)$,	$\vec{t}_{\vec{n}} = (-p + \rho g x_2, 0, 0)$
$x_2 = 0$	$\vec{n} = (0, -1, 0)$,	$\vec{t}_{\vec{n}} = (0, p, 0)$
$x_2 = \text{b}$	$\vec{n} = (0, 1, 0)$,	$\vec{t}_{\vec{n}} = (0, -p + \rho g b, 0)$
$x_3 = 0$	$\vec{n} = (0, 0, -1)$,	$\vec{t}_{\vec{n}} = (0, 0, p - \rho g x_2)$
$x_3 = \text{c}$	$\vec{n} = (0, 0, 1)$,	$\vec{t}_{\vec{n}} = (0, 0, -p + \rho g x_2)$

Pictorally,

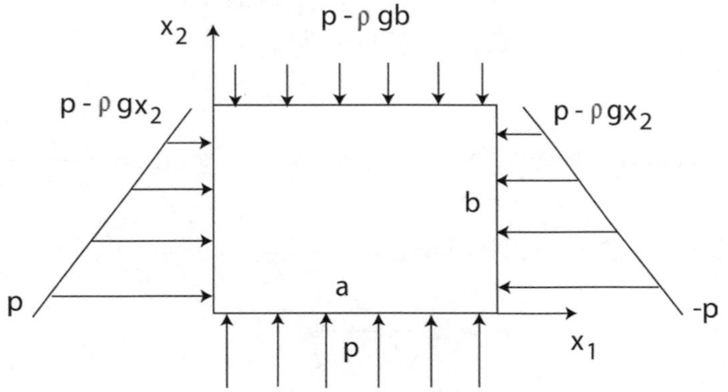

FIGURE 3.8: Pressure distribution on faces of rectangular block.

(b) On $x_2 = 0$, the total force is $\vec{F}_{x_2=0} = \int \vec{t}_{\vec{n}} \mathrm{d}A = \left(p \int \mathrm{d}A\right) \vec{e}_2 = p(ac)\vec{e}_2$.

On $x_1 = 0$, the total force is $\vec{F}_{x_1=0} = \int \vec{t}_{\vec{n}} \mathrm{d}A = \left[\int (p - \rho g x_2)\, \mathrm{d}A\right] \vec{e}_1 =$
$\left[p \int \mathrm{d}A - \rho g \int x_2 \,\mathrm{d}A\right] \vec{e}_1$, where $\mathrm{d}A = c\mathrm{d}x_2$. Thus, $\vec{F}_{x_1=0} = pbc - \frac{1}{2}\rho g b^2 c$.

3. A state of *plane stress* is one for which $T_{13} = T_{23} = T_{33} = 0$, and T_{11}, T_{22}, and T_{12} are functions of only x_1 and x_2 (see Table 3.1).
   (a) For plane stress, find the principal values and corresponding principal directions.
   (b) Determine the maximum shear.

Solution.

(a) The stress matrix for plane stress is $\left[\underset{\sim}{T}\right] = \begin{bmatrix} T_{11}(x_1,x_2) & T_{12}(x_1,x_2) & 0 \\ T_{12}(x_1,x_2) & T_{22}(x_1,x_2) & 0 \\ 0 & 0 & 0 \end{bmatrix}$. Using Eqs. (3.4) and (3.5), $\lambda[\lambda^2-(T_{11}+T_{22})\lambda+(T_{11}T_{22}-T_{12}^2)]=0$. So, $\lambda=0$ is one eigenvalue. The remaining eigenvalues are $\lambda_{1,2} = \frac{1}{2}\left((T_{11}+T_{22}) \pm \sqrt{(T_{11}-T_{22})^2+4T_{12}^2}\right) \equiv T_{1,2}$.

The eigenvectors corresponding to these eigenvalues are given by Eqs. (1.69) and (1.70).

For $\lambda = 0$, one principal value, the principal direction is $\vec{n} = \vec{e}_3$ (because $\underset{\sim}{T}\vec{e}_3 = \lambda\vec{e}_3 = 0$).

For $\lambda = T_1$ or $\lambda = T_2$,

$$\begin{aligned} (T_{11}-\lambda)a_1 + T_{12}a_2 &= 0 \\ T_{12}a_1 + (T_{22}-\lambda)a_2 &= 0 \\ -\lambda a_3 &= 0 \end{aligned}$$

The last of the above equations implies $a_3 = 0$. Thus, $a_1 = \frac{T_{12}}{\lambda - T_{11}}a_2 = \frac{\lambda - T_{22}}{T_{12}}a_2$.

With $a_1^2 + a_2^2 = 1$, and letting $A = \frac{T_{12}}{\lambda - T_{11}} = \frac{\lambda - T_{22}}{T_{12}}$,

$$a_2^2\left(A^2+1\right) = 1 \Rightarrow \begin{aligned} a_1 &= \pm\frac{A}{\sqrt{A^2+1}} \\ a_2 &= \pm\frac{1}{\sqrt{A^2+1}} \end{aligned}.$$

(b) Because the third eigenvalue is zero (i.e., $\lambda = T_3 = 0$), the maximum shear is the greatest of $\frac{|T_1 - T_2|}{2} = \frac{1}{2}\sqrt{(T_{11}-T_{22})^2+4T_{12}^2}$, $\frac{|T_1|}{2}$, and $\frac{|T_2|}{2}$.

Note that, since T_{11}, T_{22}, and T_{12} are functions of x_1, and x_2, the principal values and directions can vary from point to point.

4. In the absence of body forces, does the stress distribution given by

$$\underset{\sim}{T}=\begin{bmatrix} x_2^2+\nu(x_1^2-x_2^2) & -2\nu x_1x_2 & 0 \\ -2\nu x_1x_2 & x_1^2+\nu(x_2^2-x_1^2) & 0 \\ 0 & 0 & \nu(x_1^2+x_2^2) \end{bmatrix}$$

satisfy the equations of equilibrium?

Solution. From Eq. (3.22), $\dfrac{\partial T_{ij}}{\partial x_j}+\rho B_i=0$. With $\vec{B}=0 \Rightarrow \dfrac{\partial T_{ij}}{\partial x_j}=0$. Now,

$$\frac{\partial T_{1j}}{\partial x_j}=\frac{\partial T_{11}}{\partial x_1}+\frac{\partial T_{12}}{\partial x_2}+\frac{\partial T_{13}}{\partial x_3}=2\nu x_1-2\nu x_1+0=0$$

$$\frac{\partial T_{2j}}{\partial x_j}=\frac{\partial T_{21}}{\partial x_1}+\frac{\partial T_{22}}{\partial x_2}+\frac{\partial T_{23}}{\partial x_3}=-2\nu x_2+2\nu x_2+0=0$$

$$\frac{\partial T_{3j}}{\partial x_j}=\frac{\partial T_{31}}{\partial x_1}+\frac{\partial T_{32}}{\partial x_2}+\frac{\partial T_{33}}{\partial x_3}=0+0+0=0$$

So, yes this stress distribution satisfies the equations of equilibrium.

3.10 PROBLEMS

1. For any stress state $\underset{\sim}{T}$, the deviatoric stress, $\underset{\sim}{T}^{0}$, is defined by $\underset{\sim}{T}^{0}=\underset{\sim}{T}-\dfrac{1}{3}T_{ii}\underset{\sim}{I}$.
   (a) Show that the first invariant of the deviatoric stress vanishes.
   (b) Find a relationship between the principal values (eigenvalues) of $\underset{\sim}{T}$ and $\underset{\sim}{T}^{0}$.
   (c) Show that, in the general case, the principal directions of the stress and the deviatoric stress coincide.
2. The state of stress at a point is given by $\left[\underset{\sim}{T}\right]=\begin{bmatrix} 4 & -6 & 4 \\ -6 & 8 & 0 \\ 4 & 0 & -6 \end{bmatrix}$ MPa.
   (a) Find the stress vector at this point acting on the plane that intersects the coordinate axes at (1,0,0), (0,1,0), and (0,0,1).
   (b) Determine the magnitudes of the normal and shear stresses on this plane.
3. Regarding the stress vector (i.e., traction vector)
   (a) Given a general stress tensor, $\underset{\sim}{T}$, show that $|\vec{T}_n|=T_{ij}n_in_j$, where n_i are the components of a normal vector, $\vec{n}$, and $|\vec{T}_n|$ is the magnitude of the normal component of the stress vector (i.e., the component of the stress vector in the direction of $\vec{n}$).

(b) Using the expression derived in part (a), solve for $|\vec{T}_n|$ given the following tensor, $\underset{\sim}{T}$, and unit normal vector, $\vec{n}$.

$$\left[\underset{\sim}{T}\right] = \begin{bmatrix} 2 & 7 & 0 \\ 7 & 4 & 6 \\ 0 & 6 & -1 \end{bmatrix} \text{MPa} \qquad \vec{n} = \frac{1}{3}\begin{bmatrix} 2 \\ 2 \\ 1 \end{bmatrix}$$

(c) Find the stress vector, $\vec{t}_{\vec{n}}$, acting on this same plane.

(d) Using your answer from part (c), solve again for $|\vec{T}_n|$ to verify your answer from part (b).

4. The principal values of a stress tensor are $T_1 = 6$ MPa, $T_2 = 0.2$ MPa, and $T_3 = 10$ MPa. Find the values of T_{11} and T_{33} if a matrix of the stress is given by $\left[\underset{\sim}{T}\right] = \begin{bmatrix} T_{11} & 2 & 0 \\ 2 & 1 & 0 \\ 0 & 0 & T_{33} \end{bmatrix}$ MPa.

5. The stress tensor at a point is given by $\left[\underset{\sim}{T}\right] = \begin{bmatrix} 4 & 0 & 0 \\ 0 & 5 & 12 \\ 0 & 12 & -5 \end{bmatrix}$ MPa.

The eigenvalues and corresponding eigenvectors of this matrix are

$\lambda_1 = 4 \qquad \vec{n}_1 = \pm\vec{e}_1$

$\lambda_2 = 13 \qquad \vec{n}_2 = \pm\left(0.83205\,\vec{e}_2 + 0.5547\,\vec{e}_3\right)$

$\lambda_3 = -13 \qquad \vec{n}_3 = \pm\left(0.5547\,\vec{e}_2 - 0.83205\,\vec{e}_3\right)$

(a) What is the maximum value of *normal* stress at this point?

(b) Find the plane corresponding to the maximum normal stress (i.e., find its unit normal, $\vec{n}$).

(c) What is the maximum value of *shear* stress at this point?

(d) Find the plane corresponding to the maximum shear stress (i.e., find its unit normal, $\vec{n}$).

(e) What is the *normal stress vector* on the plane given by the unit normal $\vec{n} = \frac{2}{3}\vec{e}_1 + \frac{1}{3}\vec{e}_2 + \frac{2}{3}\vec{e}_3$?

6. The stress tensor at a given point is $\left[\underset{\sim}{T}\right] = \begin{bmatrix} 5 & 0 & -6 \\ 0 & 3 & 0 \\ -6 & 0 & -1 \end{bmatrix}$ MPa.

(a) What is the maximum value of *normal* stress among all planes passing through this point?

(b) What is the maximum value of *shear* stress among all planes passing through this point?

(c) What is the normal stress on the plane of maximum shear stress?

(d) Find the plane corresponding to the maximum normal stress (i.e., find its unit normal, $\vec{n}$).

(e) What is the shear stress on the plane from part (d)?

7. A stress field is given by $\left[\underset{\sim}{T}\right] = \begin{bmatrix} -3x_2 & x_1 & 0 \\ x_1 & x_2 & 0 \\ 0 & 0 & x_3 \end{bmatrix}$ MPa.

(a) What is the stress at the point (–3, –2, 1)?

(b) What is the max value of *normal* stress among all planes passing through this point?

(c) What is the max value of *shear* stress among all planes passing through this point?

(d) What is the normal stress on the plane of maximum shear stress?

(e) Find the plane corresponding to the max normal stress (i.e., find its unit normal, $\vec{n}$).

(f) What is the shear stress on the plane from part (e)?

8. Simple shear is the state of stress in which the only nonvanishing components are a single pair of shear stresses. If shear is occurring in the x_1 direction on planes of constant x_2, then $T_{12} = T_{21} = \tau$, and all other $T_{ij} = 0$.

(a) Find the principal values and principal directions for simple shear.

(b) Find the max shear stress and the plane on which it acts.

(c) Calculate the normal stress on the plane of maximum shear by $|\vec{T}_n|^2 = |\vec{t}_{\vec{n}}|^2 - |\vec{T}_S|^2$ and show this equals $\frac{1}{2}\left(T^{\max}_{\text{Principal}} + T^{\min}_{\text{Principal}}\right)$.

9. For the stress state of uniform compression or tension given in Table 3.1, find the principal stresses and the principal directions of these stresses. Find the maximum shear stress and the plane on which it acts.

10. In this problem we will more formally investigate the fact that the stress tensor is symmetric. Newton's second law can be stated as the sum of forces equals the time rate of change in linear momentum. Similarly, the sum of moments equals the time rate of change in angular momentum. Conservation of angular momentum in a body leads to the equation

$$\frac{\mathrm{D}}{\mathrm{D}t}\iiint\limits_R \rho\,\vec{x}\times\vec{v}\,\mathrm{d}V = \iiint\limits_R \rho\,\vec{x}\times\vec{B}\,\mathrm{d}V + \iint\limits_S \vec{x}\times\vec{t}_{\vec{n}}\,\mathrm{d}A$$

Using Eq. (1.98) and the divergence theorem (which converts the surface integral into a volume integral) we get

$$\frac{\mathrm{D}}{\mathrm{D}t}\iiint\limits_R \rho\,\varepsilon_{ijk}x_jv_k\,\mathrm{d}V = \iiint\limits_R \rho\,\varepsilon_{ijk}x_jB_k\,\mathrm{d}V + \iiint\limits_R \varepsilon_{ijk}\frac{\partial(x_jT_{pk})}{\partial x_p}\,\mathrm{d}V$$

This result must be true for all regions in the body, so the integrals can be dropped. Hence, we finally have

$$\rho\,\varepsilon_{ijk}\frac{\mathrm{D}}{\mathrm{D}t}\left(x_j v_k\right) = \varepsilon_{ijk}\left[\rho x_j B_k + \frac{\partial\left(x_j T_{pk}\right)}{\partial x_p}\right].$$

Starting from here,

(a) Expand the derivatives to show this reduces to

$$\varepsilon_{ijk}\left\{T_{jk} + x_j\left(\frac{\partial T_{pk}}{\partial x_p} + \rho B_k - \rho a_k\right) - \rho\, v_j v_k\right\} = 0$$

(b) Use Cauchy's equation of motion and calculate $\varepsilon_{ijk} v_j v_k$ to show that $\varepsilon_{ijk} T_{jk} = 0$.

(c) Finally, show that this implies $T_{ij} = T_{ji}$.

11. Show that Eq. (3.30) implies Eq. (3.21). Eq. (3.32) and Chapter 2, problem 10 will help.
12. Write the equations of motion if the stress components have the form $T_{ij} = (-p + \rho g x_2)\delta_{ij}$ where $p = p(x_1, x_2, x_3, t)$. Do NOT assume acceleration or body forces equal zero. Give your answer in both invariant and component forms.
13. Given the stress distribution for $\underset{\sim}{T}$ below, find $T_{12}(x_1, x_2)$ in order for the stress distribution to be in equilibrium given that the stress vector on the plane $x_1 = 1$ is $\vec{t}_{x_1=1} = x_2\vec{e}_1 + \left(5 - x_2^2\right)\vec{e}_2$. Assume no body forces.

$$\left[\underset{\sim}{T}\right] = \begin{bmatrix} 2x_1x_2 - x_2 & T_{12}\left(x_1, x_2\right) & 0 \\ T_{12}\left(x_1, x_2\right) & x_1^2 - x_2 & 0 \\ 0 & 0 & x_2 \end{bmatrix}$$

14. A body of density ρ is subject to body forces described by $\vec{B} = -g\vec{e}_2$ and a stress distribution as follows:

$$T_{11} = (x_1 + x_2) \qquad T_{12} = x_2(T_{11} - x_2)$$

$$T_{22} = T_{12}\frac{x_2}{x_1} \qquad T_{13} = T_{23} = 0$$

$$T_{33} = T_{12}\frac{x_3}{x_1}$$

Find the magnitude of the acceleration field within the body as a function of x_1 and x_2. Write the vector, $\vec{a}$, at the point (1,1,0).

• • • •

CHAPTER 4
Elasticity

4.1 SUMMARY UP TO NOW

So far we have studied the fundamentals of continuum mechanics, including kinematics of deformation, strain, and stress. In addition, we have developed three important principles of continuum mechanics:

- Continuity equation (Conservation of mass)

$$\rho \text{ div } \vec{v} + \frac{D\rho}{Dt} = 0$$

- Symmetry of stress tensor (Balance of moment of momentum)

$$\underset{\sim}{T} = \underset{\sim}{T}^T$$

- Equations of motion (Conservation of linear momentum)

$$\text{div}(\underset{\sim}{T}) + \rho\vec{B} = \rho\vec{a}$$

We will now begin to investigate how to use these equations to solve problems in continuum mechanics. To do so, we will need to relate the stress in the body to some measure of deformation, which will be accomplished through the introduction of constitutive equations. Thus far, we have studied concepts that apply to materials in general. Now, we will examine specific principles governing the behavior of different types of materials. We will begin with the most basic of models describing the linear elastic behavior of a real solid under infinitesimal deformations. Such material is described by a linear stress–strain relationship, called the *constitutive equation for a linear elastic solid.*

However, before we get to linear elasticity of infinitesimal deformations, we will look at general elasticity, focusing on hyperelasticity. We will then apply simplifying assumptions to the general theory to arrive at the linear, infinitesimal theory. We do this to highlight the assumptions that are inherent in linear elasticity. Furthermore, the constitutive equations we will study in the remaining chapters are also derived from more general theories. The more general theories are, by necessity, more mathematically complicated and beyond the scope of this introductory text.

*4.2 GENERAL ELASTICITY

An elastic material is a material for which the current stress in the body is solely determined by the current configuration of the body; the stress does not depend on how the current configuration came to be, such as the rate at which the deformation took place or other aspects of the deformation history. These latter considerations are associated with fluids and viscoelastic materials, as will be seen in subsequent chapters. Mathematically, the constitutive equation for general elastic materials relates the stress to the deformation gradient, Eq. (2.15). This can be expressed functionally as

$$\underset{\sim}{T} = \underset{\sim}{T}(\underset{\sim}{F}) \tag{4.1}$$

where $\underset{\sim}{T}$ is the Cauchy stress tensor. Based on the principle of material objectivity (or material frame indifference), which says that the stress in the body should be the same regardless of the reference frame from which it is measured, it can be shown that Eq. (4.1) must depend on $\underset{\sim}{F}$ through the left Cauchy–Green deformation tensor, $\underset{\sim}{B}$, or, equivalently, Ξ, the Eulerian strain tensor. Recall Eqs. (2.17) and (2.18) which define $\underset{\sim}{B}$ and Ξ, respectively. Thus, Eq. (4.1) becomes

$$\underset{\sim}{T} = \underset{\sim}{T}(\underset{\sim}{B}) \tag{4.2}$$

We will restrict our discussion to isotropic materials. In an isotropic material, the rotation of a particle has no influence on the stress tensor. With this in mind, the representation theorem [3] gives the most general functional representation of $\underset{\sim}{T}$ as

$$\underset{\sim}{T}(\underset{\sim}{B}) = \alpha_0 \underset{\sim}{I} + \alpha_1 \underset{\sim}{B} + \alpha_2 \underset{\sim}{B}^2 \tag{4.3}$$

where the α_i's are real functions of the scalar invariants of $\underset{\sim}{B}$

$$\alpha_i = \alpha_i(I_1, I_2, I_3) \tag{4.4}$$

Recall Eqs. (1.78), (1.79), and (1.80), or Eqs. (1.81), which define the scalar invariants of a second-order tensor. Explicitly combining Eqs. (4.3) and (4.4) yields

$$\underset{\sim}{T}(\underset{\sim}{B}) = \alpha_0(I_1, I_2, I_3)\underset{\sim}{I} + \alpha_1(I_1, I_2, I_3)\underset{\sim}{B} + \alpha_2(I_1, I_2, I_3)\underset{\sim}{B}^2 \tag{4.5}$$

In terms of the second Piola–Kirchoff stress tensor, Eq. (3.31), isotropy and material objectivity yield $\underset{\sim}{P}_2$ as the following function of $\underset{\sim}{C}$, (the right Cauchy–Green deformation tensor, Eq. (2.16))

$$\underset{\sim}{P}_2(\underset{\sim}{C}) = \beta_0(I_1, I_2, I_3)\underset{\sim}{I} + \beta_1(I_1, I_2, I_3)\underset{\sim}{C} + \beta_2(I_1, I_2, I_3)\underset{\sim}{C}^2 \tag{4.6}$$

where I_1, I_2, I_3 are now the scalar invariants of $\underset{\sim}{C}$. In passing, we note that the scalar invariants of $\underset{\sim}{B}$ and $\underset{\sim}{C}$ are identical, but the functions α and β are, in general, different. Eq. (4.5) or Eq. (4.6) are the constitutive equations of general isotropic elasticity. What remains is to determine the coefficient functions, α or β. We will now turn to a specific subset of elastic materials commonly employed in biomechanics, so-called hyperelastic materials, which will help us to determine the coefficient functions.

4.2.1 Hyperelasticity

The basic concept of hyperelasticity is that the material behaves elastically (i.e., recovers original shape) even at large deformations. It does this by storing the energy used to deform it as strain energy, which is released upon release of the applied load. This is in contrast to plasticity at large or small deformations, in which the original shape is not recovered. Hyperelastic materials are defined by a Helmholtz free-energy function, ψ, which describes how the strain energy is stored. Hyperelastic materials are a specific subset of elastic materials, but not all elastic materials are hyperelastic.

As mentioned, the stress–strain relationship for a hyperelastic material is based on a strain energy function. Based on the principle of material objectivity, the strain energy function is expressed as a function of $\underset{\sim}{C}$. It can be shown that

$$\underset{\sim}{T} = 2J^{-1}\underset{\sim}{F}\frac{\partial \psi}{\partial \underset{\sim}{C}}\underset{\sim}{F}^T \tag{4.7}$$

where ψ is the energy stored (*per unit initial volume*). Recall that $J = \det \underset{\sim}{F}$.

Now, we can ensure that the constitutive equation defined by Eq. (4.3) describes an isotropic material if ψ is a function of the invariants of $\underset{\sim}{C}$, i.e.,

$$\psi = \psi(I_1, I_2, I_3) \tag{4.8}$$

Using the chain rule,

$$\frac{\partial \psi}{\partial \underset{\sim}{C}} = \frac{\partial \psi}{\partial I_1}\frac{\partial I_1}{\partial \underset{\sim}{C}} + \frac{\partial \psi}{\partial I_2}\frac{\partial I_2}{\partial \underset{\sim}{C}} + \frac{\partial \psi}{\partial I_3}\frac{\partial I_3}{\partial \underset{\sim}{C}} \tag{4.9}$$

Furthermore [4],

$$\begin{aligned}
\frac{\partial I_1}{\partial \underset{\sim}{C}} &= \underset{\sim}{I} \\
\frac{\partial I_2}{\partial \underset{\sim}{C}} &= I_1\underset{\sim}{I} - \underset{\sim}{C}^T = I_1\underset{\sim}{I} - \underset{\sim}{C} \\
\frac{\partial I_3}{\partial \underset{\sim}{C}} &= (\underset{\sim}{C}^2)^T - I_1\underset{\sim}{C}^T + I_2\underset{\sim}{I} = I_3(\underset{\sim}{C}^{-1})^T = I_3\underset{\sim}{C}^{-1}
\end{aligned} \tag{4.10}$$

so that

$$\frac{\partial \psi}{\partial \underset{\sim}{C}} = \frac{\partial \psi}{\partial I_1}\underset{\sim}{I} + \frac{\partial \psi}{\partial I_2}\left(I_1\underset{\sim}{I} - \underset{\sim}{C}\right) + \frac{\partial \psi}{\partial I_3}I_3\underset{\sim}{C}^{-1} \tag{4.11}$$

Eq. (4.7) becomes

$$\underset{\sim}{T} = 2J^{-1}\underset{\sim}{F}\left[\frac{\partial \psi}{\partial I_1}\underset{\sim}{I} + \frac{\partial \psi}{\partial I_2}\left(I_1\underset{\sim}{I} - \underset{\sim}{C}\right) + \frac{\partial \psi}{\partial I_3}I_3\underset{\sim}{C}^{-1}\right]\underset{\sim}{F}^T \tag{4.12}$$

In terms of $\underset{\sim}{B}$, Eq. (4.12) takes the form

$$\underset{\sim}{T} = 2J^{-1}\left[\frac{\partial\psi}{\partial I_1}\underset{\sim}{B} + \frac{\partial\psi}{\partial I_2}\left(I_1\underset{\sim}{B} - \underset{\sim}{B}^2\right) + \frac{\partial\psi}{\partial I_3}I_3\underset{\sim}{I}\right]$$
$$= 2J^{-1}\left(\frac{\partial\psi}{\partial I_3}I_3\right)\underset{\sim}{I} + 2J^{-1}\left(\frac{\partial\psi}{\partial I_1} + \frac{\partial\psi}{\partial I_2}I_1\right)\underset{\sim}{B} - 2J^{-1}\left(\frac{\partial\psi}{\partial I_2}\right)\underset{\sim}{B}^2 \tag{4.13}$$

Comparing Eq. (4.13) to Eq. (4.5), we see $\underset{\sim}{T} = \underset{\sim}{T}(\underset{\sim}{B})$ and identify

$$\alpha_0(I_1, I_2, I_3) = 2J^{-1}\frac{\partial\psi}{\partial I_3}I_3$$
$$\alpha_1(I_1, I_2, I_3) = 2J^{-1}\left(\frac{\partial\psi}{\partial I_1} + \frac{\partial\psi}{\partial I_2}I_1\right) \tag{4.14}$$
$$\alpha_2(I_1, I_2, I_3) = -2J^{-1}\frac{\partial\psi}{\partial I_2}$$

Though we do not yet have expressions for the coefficients, we have reduced the number of unknown functions from three (α_0, α_1, α_2) to one (ψ). Furthermore, there has been ample research into different forms of ψ that one may choose to describe a tissue. For a purely incompressible material, the stress field can only be determined up to an arbitrary hydrostatic pressure, p. For incompressible materials, $I_3 = 1$ (see Chapter 2 problem 10), and ψ cannot depend on I_3. However, for reasons beyond the scope of this book, it is not sufficient to simply set $I_3 = 1$ and $\frac{\partial\psi}{\partial I_3} = 0$. If the material is incompressible, Eq. (4.13), becomes

$$\underset{\sim}{T} = -p\underset{\sim}{I} + 2\left(\frac{\partial\psi}{\partial I_1} + \frac{\partial\psi}{\partial I_2}I_1\right)\underset{\sim}{B} - 2\frac{\partial\psi}{\partial I_2}\underset{\sim}{B}^2 \tag{4.15}$$

Recall that $J = \det \underset{\sim}{F} = 1$ for incompressible materials. Table 4.1 gives some expressions for free-energy functions that have been used to describe various biological tissues.

Demonstration. Tendons can undergo finite deformations during the extension of joints. Consider the following hyperelastic energy function for an incompressible material [7]

$$\psi(I_1, I_2) = \alpha e^{\beta(I_1 - 3)} - \frac{\alpha\beta}{2}(I_2 - 3) \tag{4.16}$$

First, we wish to calculate the expression relating the Cauchy stress to the left Cauchy–Green strain tensor, and then we wish to derive an expression for the stress in the x_3 direction for the state of uniaxial tension.

TABLE 4.1: Some hyperelastic models used in biomechanics

MODEL NAME	FREE ENERGY EXPRESSION, $\psi = \psi(I_1, I_2, I_3)$
Compressible neo-Hookean	$\frac{\mu}{2}(I_1 - 3) - \mu \ln J + \frac{\lambda}{2}(\ln J)^2$
Incompressible neo-Hookean*	$\frac{\mu}{2}(I_1 - 3)$
Mooney–Rivlin	$\alpha(I_1 - 3) + \beta(I_2 - 3)$
Model of Mow and Holmes**[5]	$\frac{\gamma}{I_3^n} e^{\alpha(I_1-3)+\beta(I_2-3)}$
Model of Veronda and Westmann[6]	$\gamma\left(e^{\alpha(I_1-3)} - 1\right) + \beta(I_2 - 3) + g(I_3)$

*In this model, μ is the Lamé constant of linear elasticity (discussed below in Section 4.5)

**$n = \alpha + 2\beta$

Solution. The expression relating the Cauchy stress to the left Cauchy–Green strain tensor for an incompressible tissue is given by Eq. (4.15). Thus, we need to calculate $\frac{\partial \psi}{\partial I_1}$ and $\frac{\partial \psi}{\partial I_2}$.

$$\frac{\partial \psi}{\partial I_1} = \alpha\beta e^{\beta(I_1-3)}$$
$$\frac{\partial \psi}{\partial I_2} = -\frac{\alpha\beta}{2}$$

Substituting these expressions into Eq. (4.15) yields

$$\underset{\sim}{T} = -p\underset{\sim}{I} + \alpha\beta\left(2e^{\beta(I_1-3)} - I_1\right)\underset{\sim}{B} + \alpha\beta\underset{\sim}{B}^2$$

For uniaxial tension or unconfined compression of an isotropic incompressible tissue, $\underset{\sim}{F}$ has the following form [8]

$$\underset{\sim}{F} = \begin{bmatrix} \lambda_3^{-1/2} & 0 & 0 \\ 0 & \lambda_3^{-1/2} & 0 \\ 0 & 0 & \lambda_3 \end{bmatrix} \tag{4.17}$$

where λ_3 is the principal stretch ratio in the x_3-direction (direction coincident with load application). By direct calculation, $\underset{\sim}{B}$ and $\underset{\sim}{B}^2$ are

$$\underset{\sim}{B} = \begin{bmatrix} \lambda_3^{-1} & 0 & 0 \\ 0 & \lambda_3^{-1} & 0 \\ 0 & 0 & \lambda_3^2 \end{bmatrix} \quad \text{and} \quad \underset{\sim}{B}^2 = \begin{bmatrix} \lambda_3^{-2} & 0 & 0 \\ 0 & \lambda_3^{-2} & 0 \\ 0 & 0 & \lambda_3^4 \end{bmatrix} \tag{4.18}$$

Thus, in the x_3-direction,

$$T_{33} = -p + \alpha\beta[2e^{\beta(2\lambda_3^{-1} + \lambda_3^2 - 3)} - (2\lambda_3^{-1} + \lambda_3^2)]\lambda_3^2 + (\alpha\beta)\lambda_3^4$$

where we have made the substitution $I_1 = \text{tr}(\underset{\sim}{B}) = 2\lambda_3^{-1} + \lambda_3^2$. Examining this expression, one can appreciate the nonlinear nature of hyperelastic constitutive equations.

4.2.2 Approximations Leading to Linear Elasticity

Elasticity has two inherent forms of nonlinearity, geometrical and physical. Geometrical nonlinearity refers to whether or not the strains generated in the body are infinitesimal or finite. Physical nonlinearity refers to whether the constitutive equation relating stress to strain is of linear form.

We have thus far presented three strain tensors, the finite Eulerian strain tensor, $\underset{\sim}{\Xi}$, finite Lagrangian strain tensor, $\underset{\sim}{E}^*$, and the infinitesimal strain tensor, $\underset{\sim}{E}$ (see Eqs. (2.18), (2.23), and (2.26), respectively). Under the assumption of infinitesimal deformations, both the Eulerian and Lagrangian strain tensors reduce to the same infinitesimal strain tensor. In other words, $\underset{\sim}{\Xi} \approx \underset{\sim}{E}^* \approx \underset{\sim}{E}$ (also, $\underset{\sim}{T} \approx \underset{\sim}{P}_2$) . Through the relationship Eq. (2.18), and the infinitesimal assumption, Eq. (4.5) can be written as

$$\underset{\sim}{T}(\underset{\sim}{E}) = \gamma_0 \underset{\sim}{I} + \gamma_1 \underset{\sim}{E} + \gamma_2 \underset{\sim}{E}^2 \tag{4.19}$$

Though geometrically linear, this equation is physically nonlinear in $\underset{\sim}{E}$ and in the coefficients (they still depend on I_2 and I_3).

Now, let us examine a physically linear, but geometrically nonlinear situation. This is most easily accomplished by considering Eq. (4.6) and developing it to first order in terms of $\underset{\sim}{C}$ and its invariants. Using the result from part (b) of problem 3 in Chapter 2, Eq. (4.6) becomes

$$\begin{aligned} \underset{\sim}{P}_2(\underset{\sim}{E}^*) &= \beta_0 \underset{\sim}{I} + \beta_1(\underset{\sim}{I} + 2\underset{\sim}{E}^*) + \beta_2\left((\underset{\sim}{I} + 4\underset{\sim}{E}^* + 4(\underset{\sim}{E}^*)^2\right) \\ &= (\beta_0 + \beta_1 + \beta_2)\underset{\sim}{I} + 2(\beta_1 + 2\beta_2)\underset{\sim}{E}^* + 4\beta_2(\underset{\sim}{E}^*)^2 \end{aligned} \tag{4.20}$$

Per problem 5 of Chapter 2, developing the invariants of $\underset{\sim}{C}$ to first order in the invariants of $\underset{\sim}{E}^*$ yields $I_1^C \approx 3 + 2I_1^{E^*}$, $I_2^C \approx 3 + 4I_1^{E^*}$, and $I_3^C \approx 1 + 2I_1^{E^*}$, where the tensor to which the invariant refers is superscripted. Thus, the β coefficients become functions of only the first invariant of $\underset{\sim}{E}^*$. We also neglect the higher-order term, $(\underset{\sim}{E}^*)^2$. Assuming an unstressed reference state for the tissue, Eq. (4.20) reduces to [9]

$$\underset{\sim}{P}_2(\underset{\sim}{E}^*) = \lambda \operatorname{tr}(\underset{\sim}{E}^*)\underset{\sim}{I} + 2\mu \underset{\sim}{E}^* \tag{4.21}$$

where λ and μ are Lamé constants formed from the β coefficients. Eq. (4.21) is a linear relationship between the stress and the state of deformation, but remains geometrically nonlinear due to the presence of large strain tensors.

Finally, one can conceive of a situation which is simultaneously linear in both the geometric and physical senses. Applying geometric linearity to the physically linear Eq. (4.21), i.e., $\underset{\sim}{E}^* \approx \underset{\sim}{E}$ and $\underset{\sim}{P}_2 \approx \underset{\sim}{T}$, we get

$$\underset{\sim}{T}(\underset{\sim}{E}) = \lambda \operatorname{tr}(\underset{\sim}{E})\underset{\sim}{I} + 2\mu \underset{\sim}{E} \tag{4.22}$$

We now turn to an in depth discussion of this very equation.

4.3 EXPERIMENTAL OBSERVATIONS OF INFINITESIMAL LINEAR ELASTICITY

Consider a slender cylinder of cross-sectional area A under static tension. Figure 4.1 shows an idealized plot of applied load, P, versus extension, Δl, created during tension.

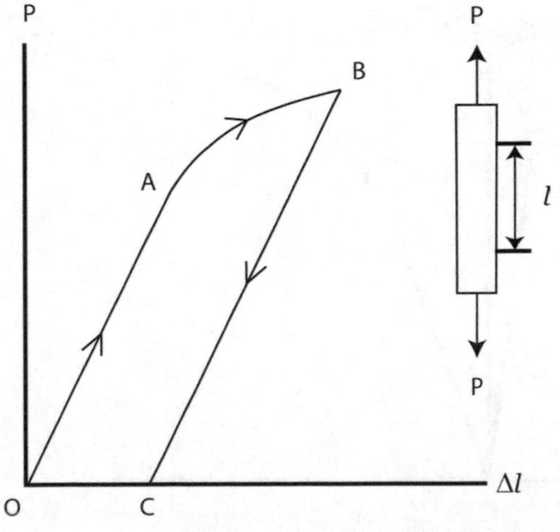

FIGURE 4.1: Load–deflection curve for a slender cylinder subjected to a uniaxial tension test.

Segment *OA* is the linear portion (proportional range); if the load is removed, line *OA* is retraced, and the specimen exhibits elasticity. Past point *A*, we traverse *OABC* and find "permanent deformation" *OC*, i.e., *plasticity*. Reapplication of the load from *C* results in elastic behavior with the same slope as *OA*, but with an increased proportional limit (material has work-hardened or been cold-worked).

The $(P, \Delta l)$ plot can be replotted as $(P/A, \Delta l/l)$ or (σ, ε) to obtain "material" (as opposed to "structural") behavior that is independent of geometry (see Figure 4.2). The axial strain is ε_a, and the lateral strain is ε_d. The slope of *OA* is a material coefficient called Young's modulus (or modulus of elasticity).

$$E_Y = \frac{\sigma}{\varepsilon_a} \tag{4.23}$$

The ratio $-\varepsilon_d/\varepsilon_a$ is another material coefficient, called the Poisson's ratio.

$$\nu = -\frac{\varepsilon_d}{\varepsilon_a} \tag{4.24}$$

If a specimen is cut at different orientations at a sufficiently small neighborhood and still exhibits the same stress–train behavior, the material is called *isotropic*. If properties change with direction, the material is *anisotropic*. If properties change at different neighborhoods, the material is *inhomogeneous* (versus homogeneous).

In addition to the stress–strain test described above, other characteristic tests can be performed as described below:

(a) To measure the change in volume of a homogeneous, isotropic material under uniform pressure p, for which the state of stress is $T_{ij} = -p\delta_{ij}$, recall the change in volume is the dilatation, $e = E_{ii}$, and define the "bulk modulus," K, as

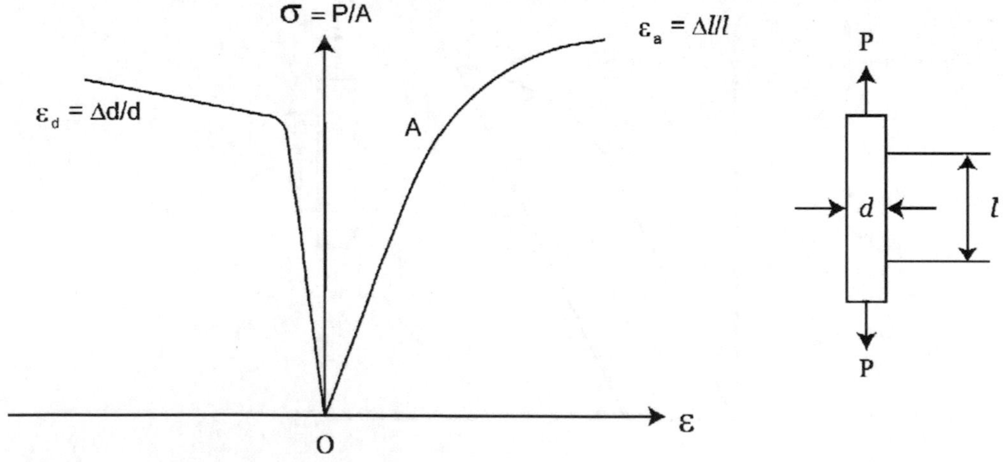

FIGURE 4.2: Stress–strain and axial versus lateral strain curves for a slender cylinder subjected to uniaxial tension.

$$K = -\frac{p}{e} \tag{4.25}$$

(b) To measure the torsional response of a cylindrical bar of circular cross-section of radius R and length l, subjected to torsional moment M_t with resulting twist angle θ, define the "shear modulus," μ, as

$$\mu = \frac{M_t l}{I_p \theta} \tag{4.26}$$

where $I_p = \dfrac{\pi R^4}{2}$ is the polar moment of inertia.

Thus, for any isotropic linear elastic material, we have introduced four material properties (E_Y, ν, K, μ) using three different experiments. But, are these constants independent? We shall see shortly. Finally, the three experiments described share the following features:

(a) relation between applied load and a quantity measuring deformation is linear

(b) rate of load application does not affect linear relationship

(c) upon removal of loading, deformations disappear completely

(d) deformations are very small

4.4 LINEARLY ELASTIC SOLID

Let us now formulate the constitutive equation of the linearly elastic, or Hookean elastic, solid. Based upon features (b), (c), and (d), we can write the basic relationship

$$\underset{\sim}{T} = \underset{\sim}{T}(\underset{\sim}{E}) \tag{4.27}$$

where $\underset{\sim}{T}$ is a single-valued function of $\underset{\sim}{E}$ with $\underset{\sim}{T}(0) = 0$. From (a), we can write

$$\begin{aligned}
T_{11} &= C_{1111}E_{11} + C_{1112}E_{12} + \ldots + C_{1133}E_{33} \\
T_{12} &= C_{1211}E_{11} + C_{1212}E_{12} + \ldots + C_{1233}E_{33} \\
&\ldots\ldots\ldots \\
T_{33} &= C_{3311}E_{11} + C_{3312}E_{12} + \ldots + C_{3333}E_{33}
\end{aligned}$$

Alternatively, these equations can be written as

$$T_{ij} = C_{ijkl}E_{kl} \tag{4.28}$$

where C_{ijkl} are the components of a fourth-order tensor ("elasticity tensor") that represents the mechanical properties of an anisotropic Hookean elastic solid. Because $\underset{\sim}{T}$ and $\underset{\sim}{E}$ are symmetric, the nine equations of Eq. (4.28) reduce to six. Thus, for a linear anisotropic elastic solid, we need no more than 36 material constants.

4.5 ISOTROPIC LINEARLY ELASTIC SOLID

If the material is isotropic, i.e., material properties are independent of direction, then $C_{ijkl} = C'_{ijkl}$ (i.e., components of elasticity tensor must not change regardless of how the bases are changed), which means the tensor is isotropic. Note that the identity tensor, $\underset{\sim}{I}$, is also isotropic, as δ_{ij} are the same for any Cartesian basis. One can also define three other isotropic fourth order tensors: $A_{ijkl} = \delta_{ij}\delta_{kl}$, $B_{ijkl} = \delta_{ik}\delta_{jl}$, and $H_{ijkl} = \delta_{il}\delta_{jk}$.

Using a theorem (beyond the scope of this book) that states that any fourth-order isotropic tensor can be represented as a linear combination of the above three tensors, we write C_{ijkl} as

$$C_{ijkl} = \lambda A_{ijkl} + \alpha B_{ijkl} + \beta H_{ijkl} \tag{4.29}$$

Moreover,

$$\begin{aligned} A_{ijkl}E_{kl} &= \delta_{ij}\delta_{kl}E_{kl} = \delta_{ij}E_{kk} = \delta_{ij}e \\ B_{ijkl}E_{kl} &= \delta_{ik}\delta_{jl}E_{kl} = E_{ij} \\ H_{ijkl}E_{kl} &= \delta_{il}\delta_{jk}E_{kl} = E_{ji} = E_{ij} \end{aligned}$$

Substitution of Eq. (4.29) into Eq. (4.28) yields

$$T_{ij} = C_{ijkl}E_{kl} = \lambda e\delta_{ij} + (\alpha + \beta)E_{ij}$$

If we let $\alpha + \beta = 2\mu$, then

$$T_{ij} = \lambda e\delta_{ij} + 2\mu E_{ij} \tag{4.30}$$

or, in invariant form,

$$\underset{\sim}{T} = \lambda e\underset{\sim}{I} + 2\mu\underset{\sim}{E} \tag{4.31}$$

Written out in long form, Eq. (4.30) becomes

$$\begin{aligned} T_{11} &= \lambda E_{kk} + 2\mu E_{11} \\ T_{22} &= \lambda E_{kk} + 2\mu E_{22} \\ T_{33} &= \lambda E_{kk} + 2\mu E_{33} \\ T_{12} &= 2\mu E_{12} \\ T_{13} &= 2\mu E_{13} \\ T_{23} &= 2\mu E_{23} \end{aligned} \tag{4.32}$$

Eqs. (4.30), (4.31), and (4.32) are different forms of the constitutive equation for a linear isotropic elastic solid. The constants, λ and μ, are called the Lamé constants (material properties). Thus, a linear isotropic elastic solid has only two independent material constants.

4.6 MATERIAL PROPERTIES OF ELASTIC MATERIALS

We will now express the material properties of a linear isotropic elastic solid (as identified in Section 4.3) using the constitutive equation, Eq. (4.30). Inverting Eq. (4.30) gives

$$E_{ij} = \frac{1}{2\mu}\left[T_{ij} - \frac{\lambda}{3\lambda + 2\mu} T_{kk}\delta_{ij}\right] \tag{4.33}$$

Also, from Eq. (4.32) or (4.33),

$$e = E_{ii} = \frac{1}{3\lambda + 2\mu} T_{kk} \tag{4.34}$$

Now, let us consider the tests we used previously to determine the properties of a linear elastic solid.

(a) Uniaxial tension (or compression). For this state of stress, only one normal stress component is nonzero. Assume $T_{11} = \sigma \neq 0$. Then, Eq. (4.33) becomes

$$\begin{aligned}
E_{11} &= \frac{1}{2\mu}\left[T_{11} - \frac{\lambda}{3\lambda + 2\mu} T_{11}\right] = \frac{\lambda + \mu}{\mu(3\lambda + 2\mu)} T_{11} \\
E_{22} = E_{33} &= -\frac{\lambda}{2\mu(3\lambda + 2\mu)} T_{11} = -\frac{\lambda}{2(\lambda + \mu)} E_{11} \\
E_{12} = E_{13} &= E_{23} = 0
\end{aligned}$$

In the tensile test from Section 4.3, we defined the Young's modulus as

$$E_Y = \frac{\sigma}{\varepsilon_a} = \frac{T_{11}}{E_{11}} = \frac{\mu(3\lambda + 2\mu)}{\lambda + \mu} \tag{4.35}$$

Similarly, the Poisson's ratio was defined as

$$\nu = -\frac{\varepsilon_d}{\varepsilon_a} = -\frac{E_{33}}{E_{11}} = \frac{\lambda}{2(\lambda + \mu)} \tag{4.36}$$

Substituting Eqs. (4.35) and (4.36) into Eq. (4.33) yields

$$\begin{aligned}
E_{11} &= \frac{1}{E_Y}[T_{11} - \nu(T_{22} + T_{33})] \\
E_{22} &= \frac{1}{E_Y}[T_{22} - \nu(T_{11} + T_{33})] \\
E_{33} &= \frac{1}{E_Y}[T_{33} - \nu(T_{11} + T_{22})] \\
E_{12} &= \frac{1}{2\mu}T_{12} \\
E_{13} &= \frac{1}{2\mu}T_{13} \\
E_{23} &= \frac{1}{2\mu}T_{23}
\end{aligned} \tag{4.37}$$

Eq. (4.37) is also referred to as the generalized Hooke's law. This is the usual form used to describe an isotropic linear elastic solid. Note that while there are three material properties, only two are independent. This is because we can show, from Eqs. (4.35) and (4.36), that

$$\mu = \frac{E_Y}{2(1+\nu)} \tag{4.38}$$

(b) In simple shear, only one pair of off-diagonal elements is nonzero. Let $T_{12} = T_{21} \neq 0$. Thus, Eq. (4.33) simplifies to

$$E_{12} = E_{21} = \frac{T_{12}}{2\mu} \tag{4.39}$$

from which we identify

$$\mu = \frac{T_{12}}{2E_{12}} \tag{4.40}$$

Thus, the Lamé constant μ is typically called the shear modulus (as identified in our torsion test).

(c) The state of hydrostatic pressure is given by $\underset{\sim}{T} = -p\underset{\sim}{I}$. Plugging this into Eq. (4.34) gives us

$$e = \frac{-3p}{2\mu + 3\lambda} \tag{4.41}$$

So, from our definition for the bulk modulus,

$$K = -\frac{p}{e} = \frac{2\mu + 3\lambda}{3} = \lambda + \frac{2}{3}\mu \tag{4.42}$$

TABLE 4.2: Relationships among different material constants used in linear elasticity

MATERIAL CONSTANT	(λ, μ)	(E_Y, ν)
λ	λ	$\dfrac{\nu E_Y}{(1+\nu)(1-2\nu)}$
μ	μ	$\dfrac{E_Y}{2(1+\nu)}$
K	$\lambda + \dfrac{2}{3}\mu$	$\dfrac{E_Y}{3(1-2\nu)}$
E_Y	$\dfrac{\mu(3\lambda+2\mu)}{\lambda+\mu}$	E_Y
ν	$\dfrac{\lambda}{2(\lambda+\mu)}$	ν

In general, we use the set (E_Y, ν) or (λ, μ). Table 4.2 provides material property relationships for an isotropic linearly elastic solid.

4.7 EQUATIONS OF THE INFINITESIMAL THEORY OF ELASTICITY

Recall the equations of motion from Chapter 3,

$$\rho a_i = \rho B_i + \frac{\partial T_{ij}}{\partial x_j}$$

We will consider only cases of small motions, where every particle is always in a small neighborhood of the natural state. In such a case, then material and spatial coordinates are approximately equal, so $a_i = \dfrac{\partial^2 u_i}{\partial t^2}$. Thus, under infinitesimal strain conditions (small deformations), we can state the equations of motion for linear isotropic elasticity as

$$\rho \frac{\partial^2 u_i}{\partial t^2} = \rho B_i + \frac{\partial T_{ij}}{\partial x_j} \tag{4.43}$$

The strain field is
$$E_{ij} = \frac{1}{2}\left(\frac{\partial u_i}{\partial x_j} + \frac{\partial u_j}{\partial x_i}\right) \tag{4.44}$$

the stress field is
$$T_{ij} = \lambda e \delta_{ij} + 2\mu E_{ij} \tag{4.45}$$

and the surface tractions are
$$t_i = T_{ij} n_j \tag{4.46}$$

Notice that Eq. (4.43) represents three equations (one for each i) for nine unknowns (three u_i and six T_{ij}). The remaining six equations come from substituting Eq. (4.44) into Eq. (4.45) (which relates the T's to the u's). Substituting that result, into Eq. (4.43) gives three equations in the three unknowns, u_1, u_2, u_3. This process yields Navier's equations of motion, which govern infinitesimal linear elasticity (see problem 4).

4.8 COMPATIBILITY CONDITIONS FOR INFINITESIMAL STRAIN CONDITIONS

When any three displacement functions u_1, u_2, u_3 are given, we can always determine the six strain components using Eq. (4.44). On the other hand, if E_{ij} are arbitrarily prescribed, it is not guaranteed that a displacement field exists that satisfies Eq. (4.44). This is because we do not necessarily obtain unique displacement components when integrating the strain components. In this case, the given E_{ij} are not compatible.

Theorem. If E_{ij} (x_1, x_2, x_3) are continuous functions, the necessary and sufficient conditions for the existence of single-valued continuous solutions $u_i(x_1, x_2, x_3)$ are

$$\begin{aligned}
\frac{\partial^2 E_{11}}{\partial x_2^2} + \frac{\partial^2 E_{22}}{\partial x_1^2} &= 2\frac{\partial^2 E_{12}}{\partial x_1 \partial x_2} \\
\frac{\partial^2 E_{22}}{\partial x_3^2} + \frac{\partial^2 E_{33}}{\partial x_2^2} &= 2\frac{\partial^2 E_{23}}{\partial x_2 \partial x_3} \\
\frac{\partial^2 E_{33}}{\partial x_1^2} + \frac{\partial^2 E_{11}}{\partial x_3^2} &= 2\frac{\partial^2 E_{31}}{\partial x_3 \partial x_1} \\
\frac{\partial^2 E_{11}}{\partial x_2 \partial x_3} &= \frac{\partial}{\partial x_1}\left(-\frac{\partial E_{23}}{\partial x_1} + \frac{\partial E_{31}}{\partial x_2} + \frac{\partial E_{12}}{\partial x_3}\right) \\
\frac{\partial^2 E_{22}}{\partial x_3 \partial x_1} &= \frac{\partial}{\partial x_2}\left(-\frac{\partial E_{31}}{\partial x_2} + \frac{\partial E_{12}}{\partial x_3} + \frac{\partial E_{23}}{\partial x_1}\right) \\
\frac{\partial^2 E_{33}}{\partial x_1 \partial x_2} &= \frac{\partial}{\partial x_3}\left(-\frac{\partial E_{12}}{\partial x_3} + \frac{\partial E_{23}}{\partial x_1} + \frac{\partial E_{31}}{\partial x_2}\right)
\end{aligned} \tag{4.47}$$

Hence, Eqs. (4.47) are called the *equations of compatibility* (or integrability).

4.9 CLASSICAL PROBLEMS IN ELASTICITY

4.9.1 Simple Infinitesimal Extension of a Linear Elastic Solid

A cylindrical elastic bar of *arbitrary cross-section* is under the action of equal and opposite normal traction, σ, at its end faces. The lateral surface is free of any surface traction. Body forces are zero, and the bar is at rest.

By considering points on the boundary surface, the stress components are

$$\begin{aligned} T_{11} &= \sigma \\ T_{22} &= T_{33} = T_{12} = T_{13} = T_{23} = 0 \end{aligned} \tag{4.48}$$

Let us show that this is a possible solution.

1. Equations of equilibrium, Eq. (4.43),

$$\frac{\partial T_{ij}}{\partial x_j} = 0$$

are identically satisfied.

2. Boundary condition on the end faces is obviously satisfied. On the lateral surface, $\vec{n} = 0\vec{e}_1 + n_2\vec{e}_2 + n_3\vec{e}_3$. From Eq. (4.46) (recall Eq. (1.32)), $\vec{t}_n = \underset{\sim}{T}\vec{n} = n_2(\underset{\sim}{T}\vec{e}_2) + n_3(\underset{\sim}{T}\vec{e}_3) = \vec{0}$. Thus, the traction-free lateral surface condition is satisfied.

3. Strain components from Eq. (4.37) are

$$\begin{aligned} E_{11} &= \frac{1}{E_Y}[T_{11} - \nu(T_{22} + T_{33})] = \frac{\sigma}{E_Y} \\ E_{22} &= \frac{1}{E_Y}[T_{22} - \nu(T_{11} + T_{33})] = -\nu\frac{\sigma}{E_Y} \\ E_{33} &= \frac{1}{E_Y}[T_{33} - \nu(T_{11} + T_{22})] = -\nu\frac{\sigma}{E_Y} \\ E_{12} &= \frac{1}{2\mu}T_{12} = 0 \\ E_{13} &= \frac{1}{2\mu}T_{13} = 0 \\ E_{23} &= \frac{1}{2\mu}T_{23} = 0 \end{aligned} \tag{4.49}$$

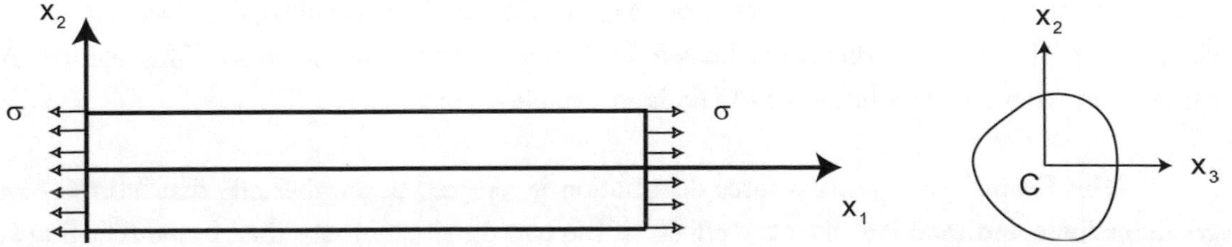

FIGURE 4.3: Longitudinal and axial cross-sections of a cylindrical elastic bar in uniaxial tension.

The conditions of compatibility, Eqs. (4.47), are automatically satisfied, so there are single-valued displacement field components.

$$\begin{aligned} u_1 &= \frac{\sigma}{E_Y} x_1 \\ u_2 &= -\frac{\nu\sigma}{E_Y} x_2 \\ u_3 &= -\frac{\nu\sigma}{E_Y} x_3 \end{aligned} \tag{4.50}$$

Thus, we have completed the solution of the simple extension or compression problem.

Assume now that the cross-sectional area of the bar is A. The surface traction σ on either side gives rise to a resultant force

$$F = \sigma A \tag{4.51}$$

passing through the centroid, C. Thus,

$$[\underset{\sim}{T}] = \begin{bmatrix} F/A & 0 & 0 \\ 0 & 0 & 0 \\ 0 & 0 & 0 \end{bmatrix} \tag{4.52}$$

Because $\underset{\sim}{T}$ is diagonal, the principal stresses are F/A, 0, 0. Hence, the maximum normal stress is $|\vec{T}_n|^{\max} = F/A$ (acting on the cross-sectional area), and the maximum shear stress is $|\vec{T}_s|^{\max} = F/2A$ (acting on planes at 45° angle to the normal cross-sections). Let L_0 be the undeformed length of the bar and Δl equal the elongation. From Eqs. (4.50) and (4.51), we can show

$$\Delta l = \frac{FL_0}{AE_Y} \tag{4.53}$$

Let d be the undeformed length of a line in the transverse direction. The contraction in the transverse direction, from Eqs. (4.50) and (4.51), is

$$\Delta d = -\nu \frac{Fd}{AE_Y} \tag{4.54}$$

It is important to note that, in practice, when a bar or similar object is pulled, the exact value of the distribution of surface traction is not known. Only the resultant force is known. This leads to the question of whether the solution we obtain is acceptable or not.

Saint Venant's principle. If a force distribution is replaced by another one that has the same resultant force and moment, then the effects of the two distributions are the same sufficiently far removed from the region of force application.

Let us briefly consider an alternative approach to this problem. In problem 4 at the end of this chapter, you will derive Navier's equations for infinitesimal elasticity. Under quasi-static conditions (i.e., ignoring inertial forces), those equations are

$$(\lambda + \mu)\frac{\partial^2 u_k}{\partial x_i \partial x_k} + \mu \frac{\partial^2 u_i}{\partial x_k^2} = 0 \tag{4.55}$$

Referring to Table 2.1, the displacements ($u_i = x_i - a_i$) associated with simple extension can be found. In each direction, the displacements are $u_i = (1 - \lambda_i)x_i$. Hence, $u_1 = u_1(x_1)$, $u_2 = u_2(x_2)$, and $u_3 = u_3(x_3)$. With this in mind, Navier's equations reduce to

$$\frac{\mathrm{d}^2 u_1}{\mathrm{d}x_1^2} = 0 \quad \frac{\mathrm{d}^2 u_2}{\mathrm{d}x_2^2} = 0 \quad \text{and} \quad \frac{\mathrm{d}^2 u_3}{\mathrm{d}x_3^2} = 0 \tag{4.56}$$

Each of these equations has the solution

$$u_i(x_i) = C_0 x_i + C_1 \tag{4.57}$$

where C_0 and C_1 are integration constants to be solved for based on the boundary conditions. We need two boundary conditions for each of u_1, u_2, and u_3. Assuming we fix one end of the bar at $x_1 = 0$, then the first boundary condition on u_1 is $u_1(x_1 = 0) = 0$. By symmetry arguments, the boundary conditions on u_2 and u_3 will be identical. By the geometry of the problem, we expect $u_2(x_2 = 0) = u_3(x_3 = 0) = 0$. Applying each of these boundary conditions to Eq. (4.57) yields $C_1 = 0$. Hence, we are left with determining C_0. Referring to Table 3.1, we see that $T_{11} = \sigma$, which again leads to Eqs. (4.49). The nonzero equations are

$$\begin{aligned} E_{11} &= \frac{\mathrm{d}u_1}{\mathrm{d}x_1} = \frac{\sigma}{E_Y} \\ E_{22} = E_{33} &= \frac{\mathrm{d}u_2}{\mathrm{d}x_2} = \frac{\mathrm{d}u_3}{\mathrm{d}x_3} = -\nu\frac{\sigma}{E_Y} \end{aligned} \tag{4.58}$$

where we have changed ∂ to d, as each displacement is only a function of its respective coordinate. The first derivative of Eq. (4.57), is

$$\frac{\mathrm{d}u_i}{\mathrm{d}x_i} = C_0 \tag{4.59}$$

Thus, applying Eqs. (4.58), we again recover Eqs. (4.50). This method of solution points out three important concepts: 1. combining conservation equations with constitutive equations can reduce the number of unknowns needing to be kept track of, 2. the resulting equation (i.e., Navier's in this case) can then be solved with appropriate boundary and initial conditions, and 3. it is often necessary to translate stress boundary conditions into displacement boundary conditions and vice versa. We will see these principles put to use again in later chapters.

4.9.2 Pure Bending of a Beam

A beam is a bar acted upon by forces/moments in an axial plane, causing bending of the bar. A beam is in *pure bending,* or *simple bending,* if acted upon by end couples only (see Figure 4.4).

Define $x_1 = 0$ and $x_1 = l$ as the left and right faces of the bar, respectively. The x_1 axis passes through the centroid of the bar. To determine a linear elasticity solution, we must specify a state of stress corresponding to

(a) a traction-free lateral surface

(b) some distribution of normal surface tractions on $x_1 = 0$ and $x_1 = l$ equivalent to the following bending couples

$$\vec{M}_R = M_2\vec{e}_2 + M_3\vec{e}_3$$
$$\vec{M}_L = -\vec{M}_R$$

Note: M_1 is absent because it would result in a twisting (torsion) couple.

We assume T_{11} is the only nonzero stress component and is a function of x_i. Then, to satisfy equilibrium

$$\frac{\partial T_{11}}{\partial x_1} = 0$$

Thus,

$$T_{11} = T_{11}(x_2, x_3) \tag{4.60}$$

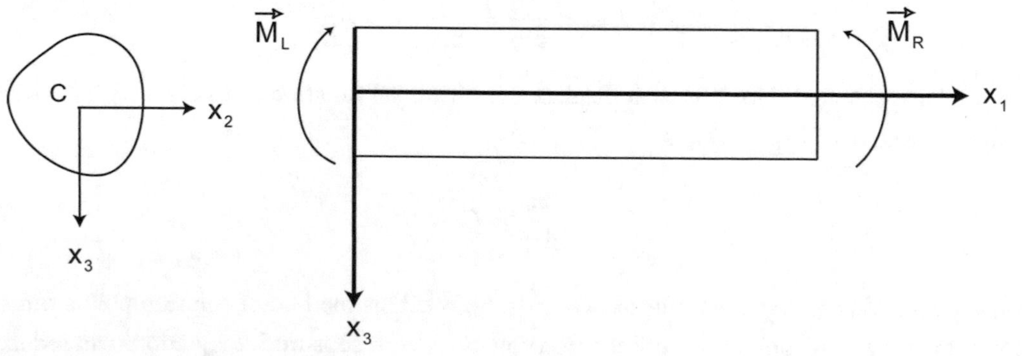

FIGURE 4.4: Longitudinal and axial cross-sections of a beam in pure bending.

and

$$E_{11} = \frac{1}{E_Y}[T_{11} - \nu(T_{22} + T_{33})] = \frac{1}{E_Y}T_{11}$$
$$E_{22} = E_{33} = -\frac{\nu}{E_Y}T_{11} \tag{4.61}$$
$$E_{12} = E_{13} = E_{23} = 0$$

Because we assumed the state of stress, let us check if the strains are compatible. Substituting Eqs. (4.61) into Eqs. (4.47), we find that $\frac{\partial^2 T_{11}}{\partial x_2^2} = 0$, $\frac{\partial^2 T_{11}}{\partial x_3^2} = 0$, and $\frac{\partial^2 T_{11}}{\partial x_2 \partial x_3} = 0$. Together these equations imply

$$T_{11} = \alpha + \beta x_2 + \gamma x_3 \tag{4.62}$$

Now, let us consider the boundary tractions. We stated that the lateral surface is traction-free. What about the two end surfaces?

$x_1 = l$: On this surface, $\vec{t}_{\vec{n}} = \underset{\sim}{T}\vec{n} = \underset{\sim}{T}\vec{e}_1 = T_{11}\vec{e}_1$. Now, the reaction force is given by $\vec{R} = \int \vec{t}_{\vec{n}}\,\mathrm{d}A$. Using Eq. (4.62),

$$R_1 = \int T_{11}\mathrm{d}A = \alpha\int \mathrm{d}A + \beta\int x_2\mathrm{d}A + \gamma\int x_3\mathrm{d}A = \alpha A \tag{4.63}$$

Note: The first moment of area about a centroidal axis is zero. Thus, $\int x_2\mathrm{d}A = \int x_3\mathrm{d}A = 0$. Similarly calculated, $R_2 = R_3 = 0$, and, from the statement of the problem, $M_1 = 0$. Now,

$$M_2 = \int x_3 T_{11}\mathrm{d}A = \alpha\int x_3\mathrm{d}A + \beta\int x_2 x_3\mathrm{d}A + \gamma\int x_3^2\mathrm{d}A = \beta I_{23} + \gamma I_{22} \tag{4.64}$$

and

$$M_3 = -\beta I_{33} - \gamma I_{23} \tag{4.65}$$

Note: The negative signs result from the right-hand rule. I_{22}, I_{33}, I_{23} are the moments and products of inertia.

$x_1 = 0$: The resultant force system is equal and opposite to that on $x_1 = l$, as the body is in equilibrium.

Because $\vec{R} = R_1\vec{e}_1$ passes through the centroid, its effect (of simple extension) can be superposed onto that of bending, i.e., $\alpha = 0$. Recall that an eccentric force (a force not coinciding with the centroid of the cross-section) is tantamount to a bending couple plus a centric force. Thus, Eq. (4.62) can be written as

$$T_{11} = \beta x_2 + \gamma x_3 \tag{4.66}$$

Without any loss of generality, we choose axes x_2 and x_3 to coincide with the principal axes of the cross-sectional area, i.e., along lines of symmetry, such that $I_{23} = 0$. By setting the axes this way, we find that Eq. (4.65) becomes

$$\beta = -\frac{M_3}{I_{33}}$$

Similarly, Eq. (4.64) becomes

$$\gamma = \frac{M_2}{I_{22}}$$

Substituting these equations into Eq. (4.66) yields the stress distribution for the cylindrical bar, with all other $T_{ij} = 0$.

$$T_{11} = \frac{M_2}{I_{22}} x_3 - \frac{M_3}{I_{33}} x_2 \tag{4.67}$$

Let us now describe the resultant deformations. For simplicity, let $M_3 = 0$. Using the strains from Eq. (4.61) and substituting in Eq. (4.67), we find

$$\begin{aligned} E_{11} &= \frac{M_2}{I_{22} E_Y} x_3 \\ E_{22} = E_{33} &= -\frac{\nu M_2}{I_{22} E_Y} x_3 \end{aligned} \tag{4.68}$$

Integrating Eqs. (4.68), one can find the displacement components [10]

$$\begin{aligned} u_1 &= \frac{M_2}{I_{22} E_Y} x_1 x_3 \\ u_2 &= -\frac{\nu M_2}{I_{22} E_Y} x_2 x_3 \\ u_3 &= -\frac{M_2}{2 I_{22} E_Y} \left[x_1^2 - \nu \left(x_2^2 - x_3^2 \right) \right] \end{aligned} \tag{4.69}$$

Consider displacements for $x_1 = \text{constant}$ (cross-sectional planes). Then,

$$u_1 = \left(\frac{M_2 x_1}{E_Y I_{22}} \right) x_3 \tag{4.70}$$

Hence, u_1 is a linear function of x_3, i.e., the cross-sectional plane remains plane and is rotated by an angle θ.

The displacement of the neutral axis is used to define the deflection of the beam as shown in Figure 4.5. For small θ,

$$\theta \approx \tan\theta = \frac{u_1}{x_3} = \frac{M_2 x_1}{E_Y I_{22}}$$

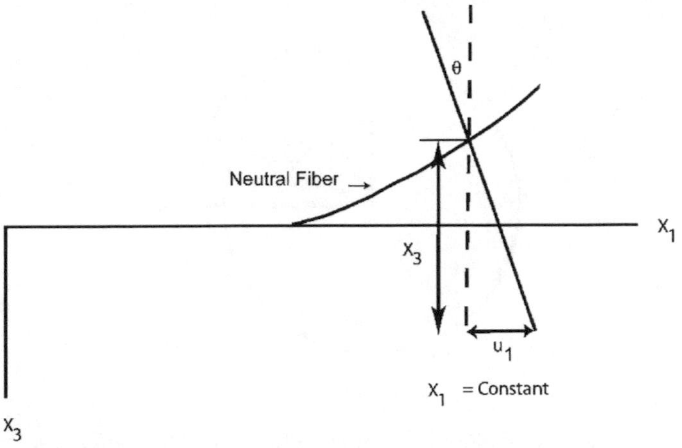

FIGURE 4.5: Displacement of the neutral axis in a beam subjected to pure bending.

The absence of shear in pure bending means that the cross-sectional planes remain perpendicular to the neutral axis. Therefore, the neutral axis experiences a state of zero stress.

4.9.3 Torsion of a Circular Cylinder

Consider the elastic deformation of a cylindrical bar of *circular* cross-section (radius a, length l), twisted by equal and opposite moments M_t as shown in Figure 4.6. It is important to note that, *unlike the results of the previous two problems*, results of this analysis apply only to circular cross-sections.

Assume that the motion of each cross-sectional plane is a rigid body rotation about x_1, i.e., $\theta = \theta(x_1)$. θ is a small rotation angle (of each cross-sectional plane), and u_i is the displacement field associated with the rotation θ.

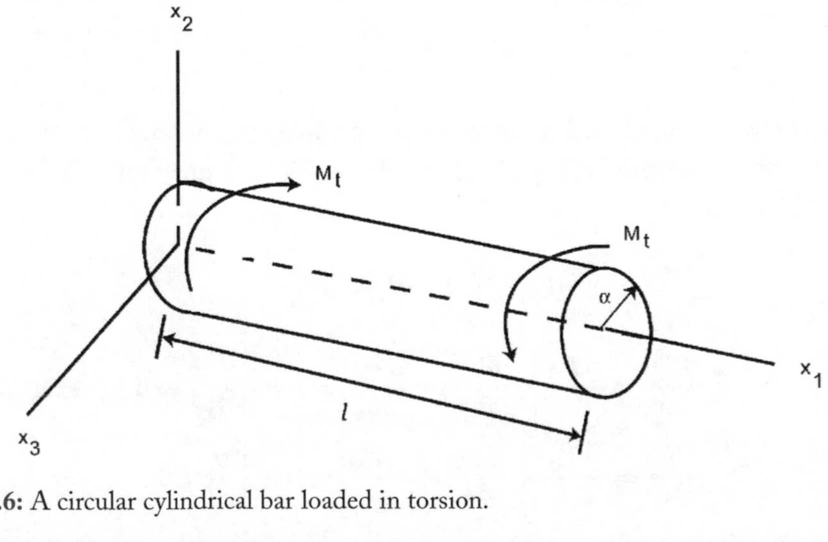

FIGURE 4.6: A circular cylindrical bar loaded in torsion.

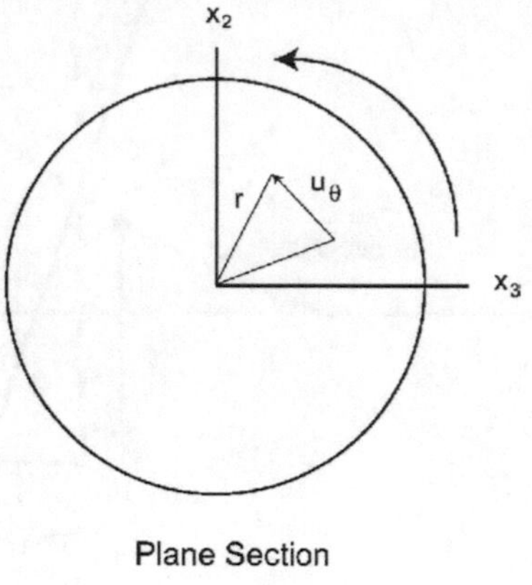

FIGURE 4.7: Circular plane cross-section of cylindrical bar loaded in torsion illustrating displacements.

The displacements are (see Figure 4.7)

$$\begin{aligned} u_1 &= 0 \\ u_2 &= -\theta x_3 \\ u_3 &= \theta x_2 \end{aligned} \tag{4.71}$$

Thus, the nonzero strains are

$$\begin{aligned} E_{12} &= -\frac{1}{2} x_3 \frac{\mathrm{d}\theta}{\mathrm{d}x_1} \\ E_{13} &= \frac{1}{2} x_2 \frac{\mathrm{d}\theta}{\mathrm{d}x_1} \end{aligned} \tag{4.72}$$

and the nonzero stresses are

$$\begin{aligned} T_{12} &= -\mu x_3 \frac{\mathrm{d}\theta}{\mathrm{d}x_1} \\ T_{13} &= \mu x_2 \frac{\mathrm{d}\theta}{\mathrm{d}x_1} \end{aligned} \tag{4.73}$$

Notice that, given $\vec{u}$, we do not need to check the compatibility condition. However, is Eq. (4.73) a possible state of stress? Substituting Eq. (4.73) into the equilibrium equation in the absence of body forces, Eq. (3.22), we find

$$\begin{aligned} \frac{\partial T_{1j}}{\partial x_j} &= \frac{\partial T_{11}}{\partial x_1} + \frac{\partial T_{12}}{\partial x_2} + \frac{\partial T_{13}}{\partial x_3} = 0 \\ \frac{\partial T_{2j}}{\partial x_j} &= \frac{\partial T_{21}}{\partial x_1} + \frac{\partial T_{22}}{\partial x_2} + \frac{\partial T_{23}}{\partial x_3} = -\mu x_3 \frac{\mathrm{d}^2\theta}{\mathrm{d}x_1^2} = 0 \\ \frac{\partial T_{3j}}{\partial x_j} &= \frac{\partial T_{31}}{\partial x_1} + \frac{\partial T_{32}}{\partial x_2} + \frac{\partial T_{33}}{\partial x_3} = \mu x_2 \frac{\mathrm{d}^2\theta}{\mathrm{d}x_1^2} = 0 \end{aligned}$$

Thus, equilibrium is satisfied *iff* the increment in angular rotation (twist per unit length) is constant, i.e.,

$$\frac{\mathrm{d}\theta}{\mathrm{d}x_1} = \theta' \tag{4.74}$$

where θ' is a constant.

Now, let us examine the surface tractions. First, the lateral surface (see Figure 4.8).

$$\vec{t}_{\vec{n}_{\text{lateral}}} = [\underset{\sim}{T}][\vec{n}_{\text{lateral}}] = \frac{1}{\alpha}\begin{bmatrix} 0 & T_{12} & T_{13} \\ T_{21} & 0 & 0 \\ T_{31} & 0 & 0 \end{bmatrix}\begin{bmatrix} 0 \\ x_2 \\ x_3 \end{bmatrix} = \frac{1}{\alpha}\begin{bmatrix} x_2T_{12} + x_3T_{13} \\ 0 \\ 0 \end{bmatrix}$$

Using Eqs. (4.73) and (4.74), $\vec{t}_{\vec{n}_{\text{lateral}}} = \frac{\mu}{\alpha}(-x_2x_3\theta' + x_2x_3\theta')\vec{e}_1 = 0 \Rightarrow$ a traction free lateral surface. Now, the end faces (see Figure 4.9). On the face $x_1 = l$, $\vec{n} = \vec{e}_1$. So,

$$\vec{t}_{\vec{n}_{\text{end}}} = \underset{\sim}{T}\vec{e}_1 = T_{21}\vec{e}_2 + T_{31}\vec{e}_3 \tag{4.75}$$

The surface traction of Eq. (4.75) gives rise to the resultant forces

$$\begin{aligned}
R_1 &= \int T_{11}\mathrm{d}A = 0 \\
R_2 &= \int T_{21}\mathrm{d}A = -\mu\theta' \int x_3\mathrm{d}A = 0 \\
R_3 &= \int T_{31}\mathrm{d}A = \mu\theta' \int x_2\mathrm{d}A = 0 \\
M_1 &= \int (x_2T_{31} - x_3T_{21})\mathrm{d}A = \mu\theta' \int \left(x_2^2 + x_3^2\right)\mathrm{d}A = \mu\theta' I_p \\
M_2 &= M_3 = 0
\end{aligned}$$

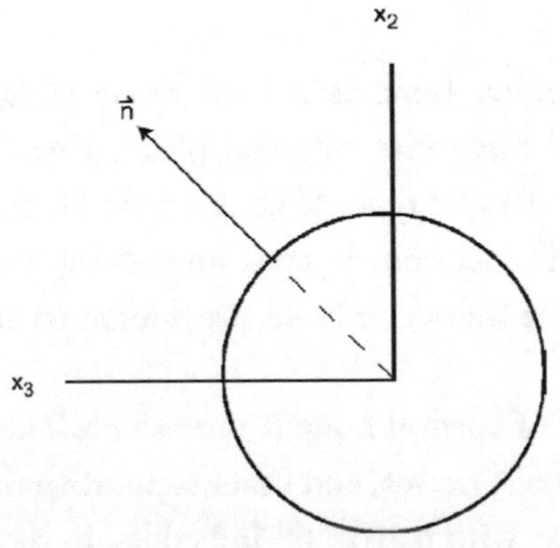

FIGURE 4.8: Surface normal defining lateral surface of circular cylindrical bar.

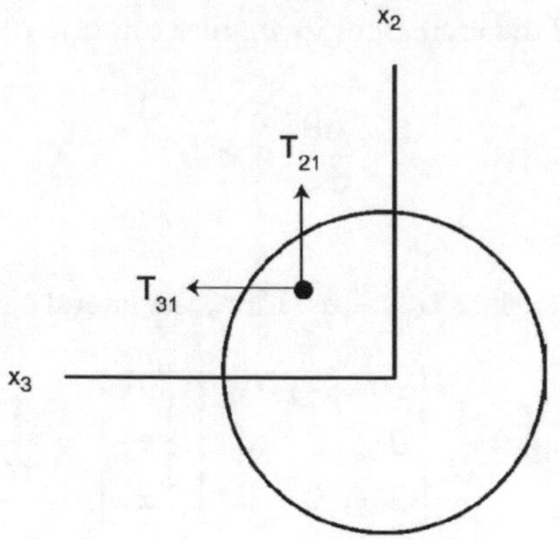

FIGURE 4.9: Shear stress components on end faces.

where $I_p = \dfrac{\pi \alpha^4}{2}$ is the polar second moment of area for a circle of radius α. On the end face $x_1 = 0$, there is a counterbalancing moment, $-\mu\theta' I_p$, to maintain equilibrium. On either end face, the resultant is a twisting moment, $M_1 = M_t$, which creates a twist per unit length given by

$$\theta' = \frac{M_t}{\mu I_p} \tag{4.76}$$

So, the stress inside the bar is

$$[\underset{\sim}{T}] = \begin{bmatrix} 0 & -\dfrac{M_t x_3}{I_p} & \dfrac{M_t x_2}{I_p} \\ -\dfrac{M_t x_3}{I_p} & 0 & 0 \\ \dfrac{M_t x_2}{I_p} & 0 & 0 \end{bmatrix} \tag{4.77}$$

Demonstration.

Torsional loading of the tibia. Bone is a hard tissue composed primarily of water, type I collagen (organic phase), and bioapatite (mineral phase, chemically similar to hydroxyapatite, $Ca_{10}(PO_4)_6(OH)_2$). There are two major types of bone: cortical (compact) bone and trabecular (cancellous or spongy) bone. We will focus our interest on cortical bone, which, even though it comprises only a shell surrounding the trabecular bone, provides most of the mechanical integrity of the long bones.

The basic structural unit of cortical bone is the osteon. The osteon consists of a Haversian canal, which contains blood vessels, nerves, and other supportive tissues, surrounded by concentric layers of mineralized bone tissue, with osteocytes imbedded in those layers (see Figure 4.10). Os-

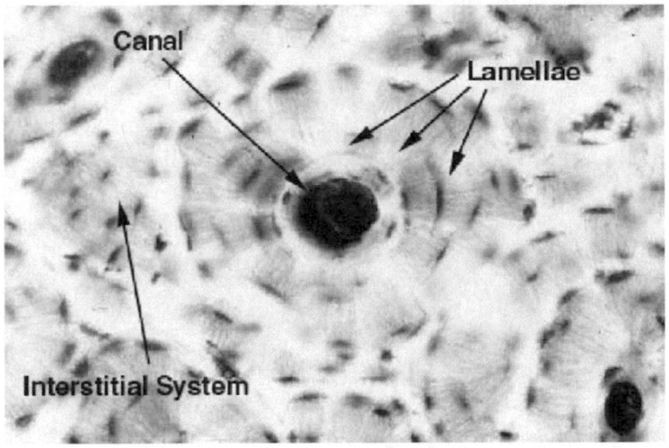

FIGURE 4.10: Micrograph of an osteon showing the Haversian canal and lamellae (concentric layers of mineralized bone).

teons (and the collagen fibers within them) are oriented in a longitudinal direction. As one might expect, this anisotropic arrangement makes bones more susceptible to damage due to transverse or shear loads. Under uniaxial tension, compact bone has a Young's modulus of 18 GPa and an ultimate tensile strength of 140 MPa. However, the shear modulus of compact bone is only 3 GPa, and the shear strength is approximately 50 MPa.

Let us use the above example for torsional loading as a starting point. Consider the lower leg of a skier, whose foot is anchored into a ski. During a skiing mishap, the skier's tibia undergoes a quasi-static twisting, with the foot locked in a stationary point as a force is applied at the end of the ski. Let us describe the tibia as a circular cylinder with radius $a = 1.5$ cm and length $l = 35$ cm. Assume that the moment generated by the skier's twisting motion is approximately 400 N·m. Will the tibia fail as a result of this accident? What is the maximum angle of twist in the bone caused by the accident?

Solution. Taking any of the stress terms from Eq. (4.77) and substituting in the appropriate values for the moment, polar moment of inertia (based on radius), and the radius, one obtains the following values for the stress and maximum angle of twist.

$$T_{13} = \frac{M_t x_2}{I_p} = \frac{2M_t \alpha}{\pi \alpha^4} = \frac{2(400)(0.015)}{\pi (0.015)^4} = 75.5 \text{ MPa}$$

$$\theta' l = \frac{M_t l}{\mu I_p} = \frac{2(400)(0.35)}{(3 \times 10^9)\pi (0.015)^4} \cong 0.587 \text{ rad} = 33.6^\circ$$

Thus, the tibia breaks (Ouch!), and the maximum angle of twist is relatively large.

4.10. PLANAR APPROXIMATIONS (2D SIMPLIFICATION)

4.10.1 Plane Strain

The state of plane strain can be defined as deformation of a cylindrical body such that there is no axial component of displacement. Let $\vec{e}_3$ be the unit vector corresponding to the cylindrical axis.

$$\begin{aligned} u_1 &= u_1(x_1, x_2) \\ u_2 &= u_2(x_1, x_2) \\ u_3 &= 0 \end{aligned} \tag{4.78}$$

Eq. (4.78) is the definition of *plane strain*. From Eq. (4.78), we can identify the nonzero strain components.

$$\begin{aligned} E_{11} &= \frac{\partial u_1}{\partial x_1} \\ E_{22} &= \frac{\partial u_2}{\partial x_2} \\ E_{12} &= \frac{1}{2}\left(\frac{\partial u_1}{\partial x_2} + \frac{\partial u_2}{\partial x_1}\right) \end{aligned} \tag{4.79}$$

Substituting Eq. (4.79) into the generalized Hooke's law, Eq. (4.32), yields

$$\begin{aligned} T_{11} &= \lambda E_{kk} + 2\mu E_{11} \\ T_{22} &= \lambda E_{kk} + 2\mu E_{22} \\ T_{33} &= \lambda E_{kk} \\ T_{12} &= 2\mu E_{12} \\ T_{23} &= T_{13} = 0 \end{aligned}$$

where $E_{kk} = \left(\frac{\partial u_1}{\partial x_1} + \frac{\partial u_2}{\partial x_2}\right)$. Because $E_{33} = 0$, the equation for E_{33} from Eq. (4.37) yields $T_{33} = \nu(T_{11} + T_{22})$.

Substituting the stresses into the equations of equilibrium gives

$$\begin{aligned} \frac{\partial T_{11}}{\partial x_1} + \frac{\partial T_{12}}{\partial x_2} &= 0 \\ \frac{\partial T_{21}}{\partial x_1} + \frac{\partial T_{22}}{\partial x_2} &= 0 \\ \frac{\partial T_{33}}{\partial x_3} &= 0 \end{aligned} \tag{4.80}$$

To reduce the number of equations from two to one (the third is already satisfied), we introduce the Airy stress function, $\phi(x_1,x_2)$, such that

$$T_{11} = \frac{\partial^2 \phi}{\partial x_2^2}$$
$$T_{22} = \frac{\partial^2 \phi}{\partial x_1^2}$$
$$T_{12} = -\frac{\partial^2 \phi}{\partial x_1 \partial x_2} \tag{4.81}$$
$$T_{33} = \nu \left(\frac{\partial^2 \phi}{\partial x_1^2} + \frac{\partial^2 \phi}{\partial x_2^2} \right)$$

For arbitrary $\phi(x_1,x_2)$, the equilibrium Eqs. (4.80) are automatically satisfied. Because the problem is cast in terms of stresses, we must check the compatibility equations. Five of six are satisfied identically, with the sixth being

$$\frac{\partial^2 E_{11}}{\partial x_2^2} + \frac{\partial^2 E_{22}}{\partial x_1^2} = 2\frac{\partial^2 E_{12}}{\partial x_1 \partial x_2} \tag{4.82}$$

Substituting Eqs. (4.81) into Eqs. (4.37) and using the results in Eq. (4.82), we find

$$\frac{\partial^4 \phi}{\partial x_1^4} + 2\frac{\partial^4 \phi}{\partial x_1^2 \partial x_2^2} + \frac{\partial^4 \phi}{\partial x_2^4} = 0 \tag{4.83}$$

Any Airy stress function $\phi(x_1,x_2)$ that satisfies Eq. (4.83) generates a possible elastic solution. The equation is known as the biharmonic equation. In particular, any third-degree polynomial (that also generates a linear stress and strain field) may be used. T_{ij}, E_{ij}, and, ultimately, u_1 and u_2 can be calculated given a ϕ. Some examples of plane strain are shown in Figure 4.11.

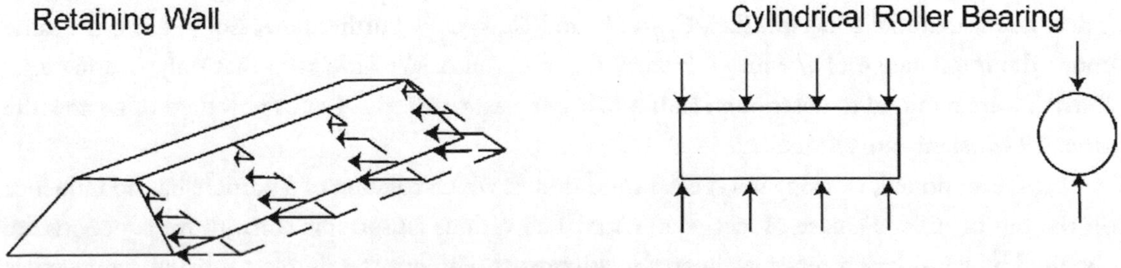

FIGURE 4.11: Examples of plane strain. For plane strain, there is no axial component of displacement.

4.10.2 Plane Stress

Plane stress is defined as $T_{33} = T_{31} = T_{32} = 0$, i.e.,

$$[\underset{\sim}{T}] = \begin{bmatrix} T_{11} & T_{12} & 0 \\ T_{12} & T_{22} & 0 \\ 0 & 0 & 0 \end{bmatrix}$$

As before, we use the Airy stress function such that

$$\begin{aligned} T_{11} &= \frac{\partial^2 \phi}{\partial x_2^2} \\ T_{22} &= \frac{\partial^2 \phi}{\partial x_1^2} \\ T_{12} &= -\frac{\partial^2 \phi}{\partial x_1 \partial x_2} \end{aligned} \tag{4.84}$$

An example of plane stress is biaxial testing of soft tissues, e.g., skin or heart valves.

Another common planar simplification of problems is to use an axis of symmetry. However, as those problems are best formulated using cylindrical coordinates, we refer the interested reader to other texts.

*4.11 ANISOTROPIC LINEAR ELASTICITY

Discussion of a general theory of anisotropic elasticity is beyond the scope of this text. However, under infinitesimal theory, anisotropy of a tissue is an easily approachable subject. We begin by recalling Eq. (4.28),

$$T_{ij} = C_{ijkl} E_{kl} \tag{4.85}$$

This equation suggests there are 81 (= 3^4) material constants needed to describe a fully anisotropic, linearly elastic material. However, considering the symmetry of both $\underset{\sim}{T}$ and $\underset{\sim}{E}$, there are only 21 independent material constants (i.e., $C_{ijkl} = C_{jikl}$ and $C_{ijkl} = C_{ijlk}$). Furthermore, Eq. (4.85) is the same under the interchange of ij with kl. Hence, $C_{ijkl} = C_{klij}$ also. We have seen that only two material constants are required to describe an isotropic linear elastic material. Let us now determine how the other 19 constants are eliminated.

As mentioned, isotropy is a special condition in which rotation of a particle has no influence on the tensor; this is a case of ideal symmetry. The various anisotropic constitutive equations are obtained by considering fewer symmetry requirements. Perhaps the simplest form of symmetry is one plane of symmetry. A material with one plane of symmetry is known as a monoclinic material.

Let $\vec{e}_1$ be normal to the symmetry plane. Then, the components of C_{ijkl} do not change under the following transformation

$$\vec{e}_1' = -\vec{e}_1 \quad \vec{e}_2' = \vec{e}_2 \quad \vec{e}_3' = \vec{e}_3 \tag{4.86}$$

where $'$ indicates the transformed coordinates. Though we have not discussed it so far, one way of defining a fourth-order tensor is that it obeys the following transformation law for an orthogonal transformation.

$$T'_{ijkl} = Q_{mi} Q_{nj} Q_{rk} Q_{sl} T_{mnrs} \tag{4.87}$$

We learned in Chapter 1 that Eqs. (4.86) define a reflection, whose transformation, $\underset{\sim}{Q}$, is

$$[\underset{\sim}{Q}] = \begin{bmatrix} -1 & 0 & 0 \\ 0 & 1 & 0 \\ 0 & 0 & 1 \end{bmatrix} \tag{4.88}$$

We see that $Q_{11} = -1$ and $Q_{22} = Q_{33} = 1$, with all other components of $\underset{\sim}{Q}$ equal to zero. Given Eqs. (4.87) and (4.88), let us consider C_{1223}. For material symmetry with respect to the above transformation, $C'_{ijkl} = C_{ijkl}$, as one should not determine any differences in material behavior between the transformed and reference coordinates. Hence, making good use of the summation convention,

$$\begin{aligned} C'_{1223} = C_{1223} &= Q_{m1} Q_{n2} Q_{r2} \left(Q_{13} C_{mnr1} + Q_{23} C_{mnr2} + Q_{33} C_{mnr3} \right) = Q_{m1} Q_{n2} Q_{r2} C_{mnr3} \\ &= Q_{m1} Q_{n2} \left(Q_{12} C_{mn13} + Q_{22} C_{mn23} + Q_{32} C_{mn33} \right) = Q_{m1} Q_{n2} C_{mn23} \\ &= Q_{m1} \left(Q_{12} C_{m123} + Q_{22} C_{m223} + Q_{32} C_{m323} \right) = Q_{m1} C_{m223} \\ &= Q_{11} C_{1223} + Q_{21} C_{2223} + Q_{31} C_{3223} = -C_{1223} \end{aligned} \tag{4.89}$$

So,

$$C_{1223} = -C_{1223} \quad \Rightarrow \quad C_{1223} = 0 \tag{4.90}$$

Similarly, $C_{1112} = C_{1113} = C_{1222} = C_{1233} = C_{1322} = C_{1323} = C_{1333} = 0$. So, we have reduced the number of constants from 21 to 13.

Now, if there are two planes of symmetry perpendicular to one another, then the third plane perpendicular to the other two is also a plane of symmetry. This transformation is given by

$$\vec{e}_1' = -\vec{e}_1 \quad \vec{e}_2' = -\vec{e}_2 \quad \vec{e}_3' = -\vec{e}_3 \tag{4.91}$$

A material with this symmetry is known as orthotropic. Given the x_2-axis is an axis of symmetry, and following what we did for the monoclinic symmetry about the x_1-axis, we additionally find that $C_{1123} = C_{2223} = C_{2333} = C_{1333} = 0$, reducing the number of material constants by four more. There is no further contribution from reflection across the x_3-axis, so there is a total of nine constants needed to define an orthotropic material. Bone is often treated as an orthotropic tissue.

Let us briefly consider how the governing equations would look for an orthotropic problem. Given the nine independent elastic constants, the six components of the stress tensor are given by the following matrix representation

$$\begin{bmatrix} T_{11} \\ T_{22} \\ T_{33} \\ T_{12} \\ T_{13} \\ T_{23} \end{bmatrix} = \begin{bmatrix} C_{1111} & C_{1122} & C_{1133} & 0 & 0 & 0 \\ C_{1122} & C_{2222} & C_{2233} & 0 & 0 & 0 \\ C_{1133} & C_{2233} & C_{3333} & 0 & 0 & 0 \\ 0 & 0 & 0 & 2C_{1212} & 0 & 0 \\ 0 & 0 & 0 & 0 & 2C_{1313} & 0 \\ 0 & 0 & 0 & 0 & 0 & 2C_{2323} \end{bmatrix} \begin{bmatrix} E_{11} \\ E_{22} \\ E_{33} \\ E_{12} \\ E_{13} \\ E_{23} \end{bmatrix} \tag{4.92}$$

where the coefficient matrix is known as the *stiffness matrix*. For $i = 1$, Eq. (4.43) is $\left(\frac{\partial T_{11}}{\partial x_1} + \frac{\partial T_{12}}{\partial x_2} + \frac{\partial T_{13}}{\partial x_3}\right) + \rho B_1 = \rho a_1$. Given Eq. (4.92), this becomes

$$\begin{aligned} &\frac{\partial}{\partial x_1}\left(C_{1111}\frac{\partial u_1}{\partial x_1} + C_{1122}\frac{\partial u_2}{\partial x_2} + C_{1133}\frac{\partial u_3}{\partial x_3}\right) \\ &+ \frac{\partial}{\partial x_2}\left(C_{1212}\left(\frac{\partial u_1}{\partial x_2} + \frac{\partial u_2}{\partial x_1}\right)\right) + \frac{\partial}{\partial x_3}\left(C_{1313}\left(\frac{\partial u_1}{\partial x_3} + \frac{\partial u_3}{\partial x_1}\right)\right) + \rho B_1 = \rho a_1 \end{aligned} \tag{4.93}$$

If we assume that the tissue is *homogeneous*, i.e., material constants are independent of location, then Eq. (4.93) becomes

$$\begin{aligned} &C_{1111}\frac{\partial^2 u_1}{\partial^2 x_1} + C_{1122}\frac{\partial^2 u_2}{\partial x_1 \partial x_2} + C_{1133}\frac{\partial^2 u_3}{\partial x_1 \partial x_3} \\ &+ C_{1212}\left(\frac{\partial^2 u_1}{\partial^2 x_2} + \frac{\partial^2 u_2}{\partial x_2 \partial x_1}\right) + C_{1313}\left(\frac{\partial^2 u_1}{\partial^2 x_3} + \frac{\partial^2 u_3}{\partial x_3 \partial x_1}\right) + \rho B_1 = \rho a_1 \end{aligned}$$

or

$$C_{1111}\frac{\partial^2 u_1}{\partial^2 x_1} + C_{1212}\frac{\partial^2 u_1}{\partial^2 x_2} + C_{1313}\frac{\partial^2 u_1}{\partial^2 x_3} + (C_{1122} + C_{1212})\frac{\partial^2 u_2}{\partial x_1 \partial x_2}$$
$$+ (C_{1133} + C_{1313})\frac{\partial^2 u_3}{\partial x_1 \partial x_3} + \rho B_1 = \rho a_1 \qquad (4.94)$$

There are similar equations for $i = 2, 3$. These last few equations are presented to demonstrate the increased complexity associated with non-isotropic problems. The material constants can be expressed in terms of the more familiar Young's moduli, Poisson's ratios, and shear moduli [2].

Moving down, a more restrictive class of symmetry is transverse isotropy. A transversely isotropic material is one where there is one plane (plane of isotropy) to which every plane perpendicular to it is also a plane of symmetry. Let $\vec{e}_3$ be the plane of isotropy. The transformation describing a transversely isotropic material is given by

$$\begin{aligned}\vec{e}_1^{\,\prime} &= (\cos\beta)\,\vec{e}_1 + (\sin\beta)\,\vec{e}_2 \\ \vec{e}_2^{\,\prime} &= (-\sin\beta)\,\vec{e}_1 + (\cos\beta)\,\vec{e}_2 \\ \vec{e}_3^{\,\prime} &= \vec{e}_3\end{aligned} \qquad (4.95)$$

More complicated mathematics show that, in addition to the constants we have already shown to be zero, Eqs. (4.95) imply that 1. $C_{1313} = C_{2323}$, 2. $C_{1133} = C_{2233}$, 3. $C_{1111} = C_{2222}$, and 4. $C_{1212} = ½(C_{1111} - C_{1122})$. Thus, the number of independent material constants is reduced from nine to five. These are commonly reported as the combination of two Young's moduli, two Poisson's ratios, and a shear modulus. Cartilage can be treated as a transversely isotropic material, where the plane of isotropy is parallel to the tissue surface.

A fully isotropic material is one in which there are two perpendicular planes of isotropy. In that case, 1. $C_{2222} = C_{3333}$, 2. $C_{1122} = C_{1133}$, and 3. $C_{1313} = C_{1212}$. So, there remain only two independent constants needed, C_{1111} and C_{1122}, to describe an isotropic linearly elastic material. With respect to the Lamé constants, $C_{1111} = \lambda + 2\mu$ and $C_{1122} = \lambda$.

4.12 PROBLEMS

1. For hydrostatic pressure, $\underset{\sim}{T} = -p\underset{\sim}{I}$. For this state of stress, find the strain tensor, $\underset{\sim}{E}$, by inverting the constitutive equation for an isotropic linear elastic solid. Use this to show the dilatation is given by $e = \dfrac{-3p}{3\lambda + 2\mu}$.

2. Using the constitutive equation of a linear isotropic elastic solid (in terms of λ and μ) and our definition of the bulk modulus (K), show that $K = \lambda + \frac{2}{3}\mu$. (Hint: Recall the experimental test used to find the bulk modulus.)
3. Consider Eq. (4.32). For an incompressible tissue, the density (or volume, given that mass is conserved) is constant. What is e for an incompressible material (recall Chapter 2, problem 11)? This must be true for any state of stress. Use the state of hydrostatic pressure to find ν for a linearly elastic, isotropic, incompressible material.
4. Navier's equations for an isotropic linear elastic solid are

$$(\lambda + \mu)\frac{\partial^2 u_k}{\partial x_i \partial x_k} + \mu \frac{\partial^2 u_i}{\partial x_k^2} = \rho \frac{\partial^2 u_i}{\partial t^2} \qquad (4.96)$$

Derive these equations using Cauchy's equation of motion and the constitutive equation for an isotropic linear elastic solid. Assume no body forces. Write Eq. 4.96 in invariant form.

5. Given that the strain tensor at a point is $[\underset{\sim}{E}] = 10^{-6}\begin{bmatrix} 25 & 30 & 0 \\ 30 & 16 & 20 \\ 0 & 20 & 9 \end{bmatrix}$

   (a) Find the stress tensor if the material properties are $\lambda = 2.4$ MPa and $\mu = 22.3$ MPa.
   (b) Find the corresponding stress vector representing this tensor on the $\vec{e}_1$ plane.
   (c) Calculate Young's modulus and Poisson's ratio.
6. Find the strain tensor corresponding to the stress tensor in Chapter 3, problem 2. Assume a linearly elastic isotropic solid with $\nu = 0.3$ and $E_Y = 4.8$ GPa.
7. Does a displacement field exist corresponding to the following strain tensor? Explain your answer.

$$[\underset{\sim}{E}] = \begin{bmatrix} 20x_1^3 & 2.5x_2^4 & 15x_1^2 \\ 2.5x_2^4 & -18x_2^8 & 7.5x_3^4 + 10x_2^3 \\ 15x_1^2 & 7.5x_3^4 + 10x_2^3 & 5x_3^4 \end{bmatrix}$$

8. A unit cube made of a linear isotropic elastic solid is subjected to uniaxial compression in the x_3-direction. Assume the cube is at equilibrium and the body force is $\vec{B} = -g\vec{e}_3$.
   (a) Calculate the stress field. (Hint: find a possible state of stress that satisfies the equilibrium equations.)
   (b) Show that a displacement field exists.

(c) Determine the surface tractions for all six faces of the cube.

9. During light exercise, the articular cartilage of the femoral head can still experience substantial stresses. Assume that the average normal compressive stress is 750 kPa, and the average stress in the lateral directions is 50 kPa. What is the percent decrease in the volume of articular cartilage? (Hint: Find Young's modulus and Poisson's ratio for hip articular cartilage in the literature, substituting aggregate modulus for Young's modulus if necessary. Please cite your source.)

10. Consider the Airy stress function $\phi = \alpha x_1^3 + \beta x_1^2 x_2 + \gamma x_1 x_2^2 + \delta x_2^3$
   (a) Does this stress function necessarily generate a possible elastic solution?
   (b) Obtain the stress tensor corresponding to this state of plane strain.
   (c) Consider a rectangular prism bounded by $x_1 = 0$, $x_1 = L$, $x_2 = \pm h/2$, $x_3 = \pm b/2$. What is the resultant force on the face $x_1 = L$?

11. Consider a cylindrical bar (with center axis aligned along x_1) made of a homogeneous, linear isotropic elastic solid. Equal and opposite normal traction σ is applied to each end of the bar. The lateral surface of the bar is free of any surface traction.
   (a) Draw a diagram of the bar. Clearly label the axes and the normal traction σ.
   (b) Define a reasonable stress field. Be sure to specify if certain components are zero.
   (c) Show that your stress field is a possible state of stress. Be sure to cover all of the important steps. State any simplifying assumptions.
   (d) Briefly explain, in words, what each condition that your stress field satisfied in part (c) means. (Hint: There are three conditions.)

12. The rectangular prismatic sample of cortical bone shown below in Figure 4.12 is placed under pure bending. A right hand couple $\vec{M} = 0.5\vec{e}_2$N·m is applied to the beam. Assume cortical bone is a linearly elastic isotropic material with $E_Y = 10$ GPa and $\mu = 4$ GPa.
   (a) Find the magnitude and location of the greatest normal stress on the beam.

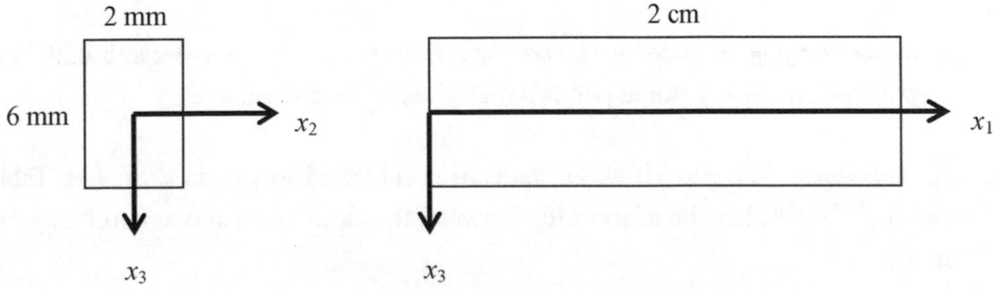

FIGURE 4.12: Rectangular sample of cortical bone for use with problem 12.

(b) Find the strain tensor at this point of greatest normal stress.

13. A circular rod of radius R is under torsion.
 (a) Find the maximum normal and shear stresses and their locations on the rod.
 (b) Find the principal strains as a function of distance along the x_2 axis, i.e., let $x_2 = r$ and $x_3 = 0$.
14. A circular cylindrical sample of cortical bone (radius = 2 mm, length = 10 mm) is subjected to a twisting moment of 1 N·m. The resulting twist at one end face is 0.15 radians. Assume cortical bone behaves like a linearly elastic isotropic solid.
 (a) Find the shear modulus of cortical bone, μ.
 (b) An identical circular cylindrical sample of cortical bone is subjected to a pure bending test, with an applied moment $M_2 = 1$ N·m. What is the deflection of the bar at $x_1 =$ 10 mm, when the other end face is located at $x_1 = 0$? Assume $\nu = 0.5$. (Hint: $I_2 = \dfrac{\pi R^4}{4}$ for a circular cross-section with a centroidal axis.)
15. Write Eq. (4.92) for a material transversely isotropic with respect to the $\vec{e}_3$ axis. What is the equation of motion for $i = 3$?
16. (a) A hyperelastic material with free energy function given by the Mow and Holmes hyperelastic model is subjected to confined compression. For confined compression in the z-direction,

$$[\underset{\sim}{F}] = \begin{bmatrix} 1 & 0 & 0 \\ 0 & 1 & 0 \\ 0 & 0 & \lambda_z \end{bmatrix} \tag{4.97}$$

 Show that the relationship between T_{zz} and λ_z is given by $T_{zz} = 2\psi\lambda_z\left(\alpha + 2\beta - \frac{n}{\lambda_z^2}\right)$.
 (b) Show that Eq. (4.7) implies

$$\underset{\sim}{T} = 2\rho\underset{\sim}{F}\frac{\partial\varphi}{\partial\underset{\sim}{C}}\underset{\sim}{F}^T \tag{4.98}$$

 where $\psi = \rho_0\varphi$. As stated in the text, ψ is the free energy stored per unit initial volume. φ is the free energy stored per unit mass, i.e., specific strain energy.

17. An incompressible neo-Hookean material is subjected to simple shear (see Tables 2.1 and 4.1). Show that the relationship between the shear stress and angle change is $T_{12} = \mu\tan\gamma$.

• • • •

CHAPTER 5

Fluids

5.1 INTRODUCTION TO FLUIDS

Fluids are a class of idealized materials that cannot support or sustain a shear stress and, as a result, conform to the boundaries of their container. Examples include water and air, which are incompressible and compressible fluids, respectively. While fluids include both gases and liquids, in this book, we are primarily interested in liquids.

In terms of stress behavior, fluids can be classified as

(a) *Newtonian* (or linearly viscous fluids)—stress depends linearly on the rate of deformation
(b) *Non-Newtonian*—e.g., polymeric solutions, blood flowing at low strain rates.

From the definition of a fluid, it follows that the stress vector on any plane at any point is normal to the plane. In other words, for any $\vec{n}$

$$\underset{\sim}{T}\vec{n} = \lambda\vec{n} \tag{5.1}$$

Because every plane is a principal plane, every direction is an eigenvector of $\underset{\sim}{T}$. Consider any two planes, $\vec{n}_1$ and $\vec{n}_2$. $\underset{\sim}{T}\vec{n}_1 = \lambda_1\vec{n}_1$, so $\vec{n}_2 \cdot \underset{\sim}{T}\vec{n}_1 = \lambda_1\vec{n}_1 \cdot \vec{n}_2$. Also, $\underset{\sim}{T}\vec{n}_2 = \lambda_2\vec{n}_2$, so $\vec{n}_1 \cdot \underset{\sim}{T}\vec{n}_2 = \lambda_2\vec{n}_1 \cdot \vec{n}_2$. These statements imply

$$\vec{n}_1 \cdot \underset{\sim}{T}\vec{n}_2 - \vec{n}_2 \cdot \underset{\sim}{T}\vec{n}_1 = (\lambda_2 - \lambda_1)\vec{n}_1 \cdot \vec{n}_2 \tag{5.2}$$

Recall Eq. (1.48) for the transpose of a tensor. Hence, we can write $\vec{n}_2 \cdot \underset{\sim}{T}\vec{n}_1 = \vec{n}_1 \cdot T^T\vec{n}_2 = \vec{n}_1 \cdot T\vec{n}_2$, as $\underset{\sim}{T}$ is symmetric. Thus, the left-hand side of Eq. (5.2) is 0. Therefore, $(\lambda_2 - \lambda_1)\vec{n}_1 \cdot \vec{n}_2 = 0 \Rightarrow \lambda_1 = \lambda_2$. What does this mean? This result means that on *all* planes through a point in a fluid, not only are there no shear stresses, the normal stresses are all the same. This is why the class of materials known as fluids is said to not be able to support shear stresses. Any shear stress applied to a fluid induces flow.

We define the normal stress as $-p$ (negative normal $\rightarrow$ compression). Thus,

$$\underset{\sim}{T} = -p\underset{\sim}{I} \tag{5.3}$$

or $T_{ij} = -p\delta_{ij}$, where p is the hydrostatic pressure.

An *incompressible fluid* is one where the density is constant with respect to time regardless of the state stress. Mathematically this is expressed as $\frac{D\rho}{Dt} = 0$. Recall the conservation of mass, Eq. (2.39),

$$\frac{D\rho}{Dt} + \rho \frac{\partial v_i}{\partial x_i} = 0 \tag{5.4}$$

So, for an incompressible fluid,

$$\frac{\partial v_i}{\partial x_i} = 0 \tag{5.5}$$

In invariant form, we have

$$\operatorname{div}(\vec{v}) = 0 \quad \text{or} \quad \nabla \cdot \vec{v} = 0 \tag{5.6}$$

Note: Incompressible fluids are not required to have a spatially uniform density (e.g., salt water with nonuniform salt concentration). If the density is uniform, it is referred to as a *homogeneous fluid* (ρ constant everywhere).

5.2 HYDROSTATICS

Recall the equations of equilibrium, Eqs. (3.22).

$$\frac{\partial T_{ij}}{\partial x_j} + \rho B_i = 0 \tag{5.7}$$

Eq. (5.3) into Eq. (5.7) yields

$$\frac{\partial p}{\partial x_i} = \rho B_i,$$

or, in invariant form,

$$\nabla p = \rho \vec{B}, \tag{5.8}$$

where B_i are components of weight per unit mass. Also, let the x_3 axis point vertically down, such that $B_1 = B_2 = 0$, and $B_3 = g$. Then, Eq. (5.8) becomes

$$\frac{\partial p}{\partial x_1} = 0, \quad \frac{\partial p}{\partial x_2} = 0, \quad \text{and} \quad \frac{\partial p}{\partial x_3} = \frac{\mathrm{d}\, p}{\mathrm{d}\, x_3} = \rho g, \tag{5.9}$$

where the partial derivative changes to a simple derivative in the last expression because the first two expressions show that p is independent of x_1 and x_2. Thus, Eq. (5.9) says that p is a function of x_3 alone; the pressure difference between two points in the liquid separated by a vertical distance h is

$$p_2 - p_1 = \rho g h \tag{5.10}$$

Thus, static pressure in the liquid depends only on the depth. It is the same for all particles on the same horizontal plane. If the fluid is in a state of rigid motion (i.e., rate of deformation = 0, but inertial effects are non-negligible), then the governing equation is

$$-\frac{\partial p}{\partial x_i} + \rho B_i = \rho a_i \tag{5.11}$$

Demonstration. A body of cross-sectional area A, length l, and weight W, is tied by a rope to the bottom of a container which is filled with a liquid (see Figure 5.1). If the density of the body, ρ, is less than that of the liquid, the body will float upward generating tension in the rope. What is the tension in the rope?

Solution. Let p_{top} and p_{bottom} be the pressure at the top and bottom surfaces of the body, respectively. Furthermore, let T be the tension in the rope. For the cylinder to be in equilibrium, the sum of the vertical forces must be zero.

$$p_{\text{bottom}}A - p_{\text{top}}A - W - T = 0 \tag{5.12}$$

So,

$$T = A(p_{\text{bottom}} - p_{\text{top}}) - W \tag{5.13}$$

But, from Eq. (5.10), $p_{\text{bottom}} - p_{\text{top}} = \rho gl$, such that

$$T = \rho glA - W \tag{5.14}$$

Note: The first term in Eq. (5.14) is the buoyancy force acting on the body.

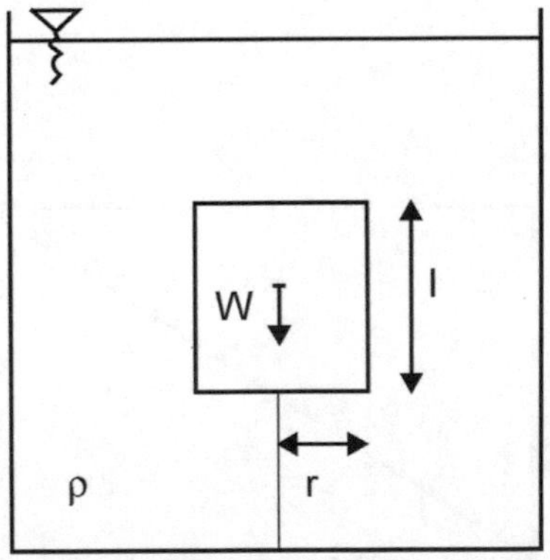

FIGURE 5.1: Example of hydrostatics. The weight of a body in a container full of liquid is balanced by the buoyant force exerted by the liquid on the body.

5.3 NEWTONIAN VISCOUS FLUID

Recall from elasticity that a solid under shear stress deforms and reaches an equilibrium state with a resultant shear strain; this is a reversible process. A fluid under shear stress deforms and reaches a steady state where it deforms continuously with a rate of shear, a process that is irreversible. Thus, for a fluid, shear stress = f(rate of shear strain). For a Newtonian fluid, this relationship is linear (see Figure 5.2).

Let us decompose $\underset{\sim}{T}$ into two parts.

$$\underset{\sim}{T} = -p\underset{\sim}{I} + \underset{\sim}{T}' \quad \text{or} \quad T_{ij} = -p\delta_{ij} + T'_{ij} \tag{5.15}$$

where $\underset{\sim}{T}'$ depends only on the rate of deformation, $\underset{\sim}{D}$. Recall the strain-rate tensor, Eq. (2.36). For a rigid body motion (i.e., $\underset{\sim}{D} = 0$), $\underset{\sim}{T}' = 0$, and the stress in the fluid is given by Eq. (5.3). However, for nonrigid body motion, following an approach similar to the isotropic linear elastic solid,

$$T'_{ij} = \lambda D_{kk}\delta_{ij} + 2\mu D_{ij} \tag{5.16}$$

where

$$D_{ij} = \frac{1}{2}\left(\frac{\partial v_i}{\partial x_j} + \frac{\partial v_j}{\partial x_i}\right)$$

The coefficients λ and μ are material constants (not the Lamé constants) and T'_{ij} is the *viscous stress tensor*. Putting Eq. (5.16) into Eq. (5.15), yields the constitutive equation for a Newtonian fluid.

$$T_{ij} = -p\delta_{ij} + \lambda D_{kk}\delta_{ij} + 2\mu D_{ij} \tag{5.17}$$

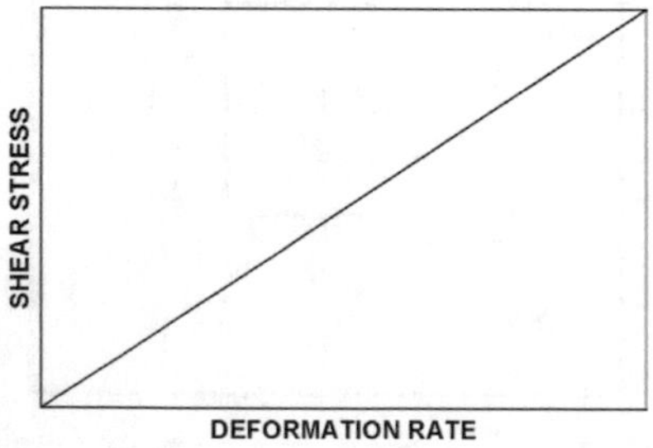

FIGURE 5.2: Relationship between shear stress and deformation rate for a Newtonian field.

where T_{ij} is the *total stress tensor*. Writing these equations out, one gets

$$
\begin{aligned}
T_{11} &= -p + \lambda D_{kk} + 2\mu D_{11} \\
T_{22} &= -p + \lambda D_{kk} + 2\mu D_{22} \\
T_{33} &= -p + \lambda D_{kk} + 2\mu D_{33} \\
T_{12} &= 2\mu D_{12} \\
T_{13} &= 2\mu D_{13} \\
T_{23} &= 2\mu D_{23}
\end{aligned}
\tag{5.18}
$$

The pressure, p, above is a little bit of ambiguous terminology because p is only part of the total compressive normal stress on a plane. Just remember that the isotropic tensor, $-p\delta_{ij}$, is the part of T_{ij} which does not depend on the rate of deformation.

5.4 MEANING OF λ AND μ

Consider the shear flow given by the velocity field

$$v_1 = v_1(x_2), \quad v_2 = v_3 = 0 \tag{5.19}$$

The strain rate tensor for this flow is

$$D_{12} = D_{21} = \frac{1}{2}\frac{\partial v_1}{\partial x_2} = \frac{1}{2}\frac{dv_1}{dx_2} \tag{5.20}$$

with all other $D_{ij} = 0$.
Substituting Eq. (5.20) into Eq. (5.18),

$$
\begin{aligned}
T_{11} &= T_{22} = T_{33} = -p \\
T_{13} &= T_{23} = 0
\end{aligned}
$$

and

$$T_{12} = \mu \frac{dv_1}{dx_2} \tag{5.21}$$

We call μ the *first coefficient of viscosity*. It has units of kg/m·s. It is a proportionality constant relating shear stress to the rate of decrease of angle between two mutually perpendicular material lines. Now, consider the trace of Eq. (5.16),

$$T'_{ii} = \lambda D_{kk}\delta_{ii} + 2\mu D_{ii} = D_{ii}(3\lambda + 2\mu)$$

Thus,

$$\frac{1}{3}T'_{ii} = \left(\lambda + \frac{2}{3}\mu\right) D_{ii} \tag{5.22}$$

$\frac{1}{3}T'_{ii}$ is the mean normal viscous stress, and D_{ii} is the rate of volume change. $\lambda + \frac{2}{3}\mu$, called the *coefficient of bulk viscosity*, is the constant of proportionality relating viscous mean normal stress to the rate of volume change.

5.5 INCOMPRESSIBLE NEWTONIAN FLUID

For an incompressible fluid, Eq. (5.5), $D_{ii} = 0$. Thus, Eq. (5.17) becomes

$$T_{ij} = -p\delta_{ij} + 2\mu D_{ij} \tag{5.23}$$

Eq. (5.23) can be written as

$$T_{ij} = -p\delta_{ij} + \mu\left(\frac{\partial v_i}{\partial x_j} + \frac{\partial v_j}{\partial x_i}\right) \tag{5.24}$$

Furthermore, from Eq. (5.23),

$$T_{ii} = -3p + 2\mu D_{ii}$$

which, as $D_{ii} = 0$ for an incompressible fluid, implies

$$p = \frac{-T_{ii}}{3} \tag{5.25}$$

Hence, for an incompressible fluid, the pressure is the mean normal compressive stress.

5.6 NAVIER–STOKES EQUATIONS

Let us substitute the constitutive equations for an incompressible viscous fluid, Eq. (5.24), into the equations of motion, Eq. (3.21). The resulting equations are the Navier–Stokes equations.

$$\rho\left(\frac{\partial v_i}{\partial t} + v_j\frac{\partial v_i}{\partial x_j}\right) = \rho B_i - \frac{\partial p}{\partial x_i} + \mu\frac{\partial^2 v_i}{\partial x_j^2} \tag{5.26}$$

where we have used $a_i = \frac{\mathrm{D}v_i}{\mathrm{D}t} = \left(\frac{\partial v_i}{\partial t} + v_j\frac{\partial v_i}{\partial x_j}\right)$. In general, for any coordinate system, $\vec{a} = \frac{\mathrm{D}\vec{v}}{\mathrm{D}t} = \frac{\partial \vec{v}}{\partial t} + (\nabla\vec{v})\vec{v}$.

Note: ∂x_j^2 is equivalent to $\partial x_j \partial x_j$, and is thus a summation. For example, the equation for $i = 1$ is

$$\rho\left(\frac{\partial v_1}{\partial t}+v_1\frac{\partial v_1}{\partial x_1}+v_2\frac{\partial v_1}{\partial x_2}+v_3\frac{\partial v_1}{\partial x_3}\right)=\rho B_1-\frac{\partial p}{\partial x_1}+\mu\left(\frac{\partial^2 v_1}{\partial x_1^2}+\frac{\partial^2 v_1}{\partial x_2^2}+\frac{\partial^2 v_1}{\partial x_3^2}\right)$$

In invariant form,

$$\rho\left(\frac{\partial \vec{v}}{\partial t}+(\nabla\vec{v})\vec{v}\right)=-\nabla p+\mu\nabla^2\vec{v}+\rho\vec{B} \tag{5.27}$$

These are the *Navier–Stokes equations of motion for an incompressible Newtonian fluid.* At physiological pressures, most biological fluids can be considered incompressible and, thus, can be analyzed with these equations. We have three equations with four unknown functions, v_1, v_2, v_3, and p. The fourth equation is supplied by the continuity equation for an incompressible fluid, Eq. (5.6)

$$\nabla\cdot\vec{v}=\frac{\partial v_1}{\partial x_1}+\frac{\partial v_2}{\partial x_2}+\frac{\partial v_3}{\partial x_3}=0 \tag{5.28}$$

Demonstration. If all particles have their velocity vectors parallel to a fixed direction, the flow is said to be parallel, or unidirectional flow. Show that for parallel flows of an incompressible, linearly viscous fluid, the total normal compressive stress at any point on any plane parallel to and perpendicular to the direction of flow is the pressure, p.

Solution. Let the x_1 axis be the flow direction ($v_2 = v_3 = 0$). From the continuity equation, $\frac{\partial v_1}{\partial x_1} = 0$. Thus, the velocity field is independent of x_1 and must be of the form $v_1 = v_1(x_2, x_3, t)$. For this flow, $D_{11} = D_{22} = D_{33} = 0$. Hence, from Eq. (5.23), $T_{11}=T_{22}=T_{33}=-p$.

5.7 BOUNDARY CONDITION

On a rigid boundary, we consider a *"no-slip" condition* exists. This means a fluid layer next to a rigid surface moves with that surface. If that surface is at rest, then the fluid velocity at that surface equals zero. This is an experimental fact based on numerous observations; a "no-slip" condition applies to practically all fluids.

5.8 IMPORTANT DEFINITIONS

- *Steady flow*—Any property may vary from point to point, but all properties are constant with respect to time at every point.

For steady flow, $\frac{\partial \rho}{\partial t} = 0$, but $\frac{D\rho}{Dt} \neq 0$ because $\rho = \rho\,(x_1, x_2, x_3)$. So, by Eq. (5.4), conservation of mass during steady flow is described by

$$\nabla \cdot \rho \vec{v} = 0$$

or, in indicial form,

$$\frac{\partial(\rho v_i)}{\partial x_i} = 0 \tag{5.29}$$

- *Streamline*—One of four basic representations of a flow field. Streamlines are drawn such that they are tangent to the instantaneous velocity of flow at every point.
- *Laminar vs. turbulent flow*—Characterization of the flow structure within viscous flow regimes.
  - Laminar regime → smooth, layered flow structure
  - Turbulent regime → random fluid particle motion in flow structure
  - The *Reynolds number* (Re) is a dimensionless parameter used to distinguish between flow structures. For internal flow in a pipe

$$\mathrm{Re} = \frac{\rho v d}{\mu} \tag{5.30}$$

Where ρ is the fluid density, v is the mean fluid velocity, d is the pipe diameter, and μ is the fluid viscosity. Re is a dimensionless parameter representing the ratio of inertial forces to viscous forces.

$$\begin{aligned} \mathrm{Re} &\leq 2{,}300 \Rightarrow \text{Laminar flow} \\ \mathrm{Re} &> 2{,}300 \Rightarrow \text{Turbulent flow } \textit{likely} \end{aligned} \tag{5.31}$$

5.9 CLASSICAL FLOWS

5.9.1 Plane Couette Flow

Plane Couette flow is steady, unidirectional flow of an incompressible viscous fluid between two horizontal planes of infinite extent with no pressure gradient in the flow direction. One plate is fixed; the other is moving with constant velocity v_0. The velocity field is, $v_1 = v_1(x_2)$ and $v_2 = v_3 = 0$. The "no-slip" boundary conditions are

$$\begin{aligned} v_1(0) &= 0 \\ v_1(d) &= v_0 \end{aligned} \tag{5.32}$$

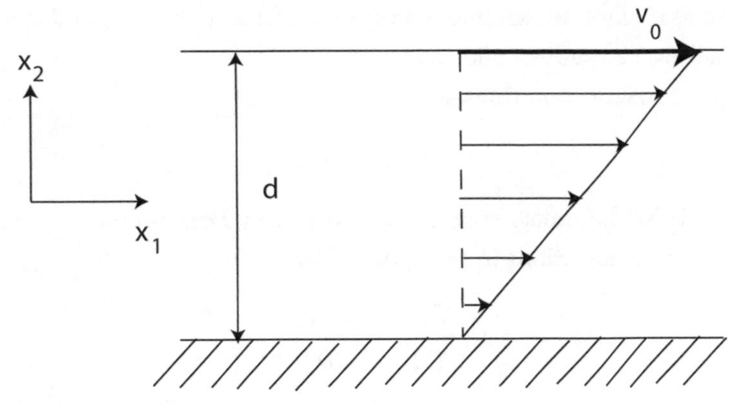

FIGURE 5.3: Plane Couette flow.

From the Navier–Stokes equations,

$$\frac{d^2 v_1}{dx_2^2} = 0 \tag{5.33}$$

Solving Eq. (5.33) subject to Eq. (5.32) gives

$$v_1(x_2) = v_o \frac{x_2}{d} \tag{5.34}$$

This is the solution for plane Couette flow. It is a linear function of the distance from the rigid boundary as shown in Figure 5.3.

5.9.2 Plane Poiseuille Flow

Plane Poiseuille flow is steady, unidirectional flow of an incompressible Newtonian viscous fluid between two fixed parallel plates of "infinite extent", i.e., the size of the channel in the x_1 and x_3 directions is much greater than the x_2 direction. Let the height of the channel be $2a$ (see Figure 5.4),

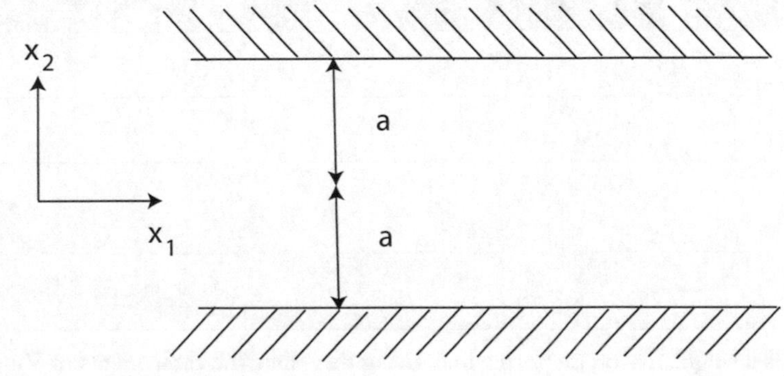

FIGURE 5.4: Geometry of Plane Poiseuille flow.

and the width be w. As before, we assume the velocity field is, $v_1 = v_1(x_2)$ and $v_2 = v_3 = 0$; but, this time, there is also a pressure gradient allowed.

The "no-slip" boundary conditions are

$$v_1(\pm a) = 0 \tag{5.35}$$

Note that Eq. (5.35) is two boundary conditions; one for a and one for $-a$.

From Navier–Stokes equations in the x_1 direction,

$$\frac{\partial p}{\partial x_1} = \mu \frac{d^2 v_1}{dx_2^2} \tag{5.36}$$

From Navier–Stokes equations in the x_2 and x_3 directions,

$$\frac{\partial p}{\partial x_2} = 0 \quad \text{and} \quad \frac{\partial p}{\partial x_3} = 0 \tag{5.37}$$

from which we identify the pressure is only a function of x_1, $p = p(x_1)$. So, Eq. (5.36) becomes

$$\frac{dp}{dx_1} = \mu \frac{d^2 v_1}{dx_2^2} \tag{5.38}$$

Differentiating Eq. (5.38) with respect to x_1, we get

$$\frac{d^2 p}{dx_1^2} = 0$$

which means $\frac{dp}{dx_1}$ = constant; i.e., the pressure gradient is a constant along the flow direction. Treating $\frac{dp}{dx_1}$ as a constant, and integrating Eq. (5.38) twice with respect to x_2, we obtain

TABLE 5.1: Some basic flow patterns

FLOW PATTERN	v_1	v_2	v_3
Simple shearing	cx_2	0	0
Rectilinear	$v_1(x_1, x_2, x_3)$	0	0
Vortex*	0	$\frac{-cx_3}{x_2^2 + x_3^2}$	$\frac{cx_2}{x_2^2 + x_3^2}$
Plane	0	$v_2(x_2, x_3, t)$	$v_3(x_2, x_3, t)$

*$x_2^2 + x_3^2 \neq 0$. There is a singularity on the vortex line. Using this table, the components of $\nabla\vec{v}$ and the strain rate tensor can be found.

$$\mu v_1 = \left(\frac{dp}{dx_1}\right)\frac{x_2^2}{2} + C_1x_2 + C_2 \tag{5.39}$$

Applying the boundary conditions, Eq. (5.35), to Eq. (5.39) yields

$$v_1 = \frac{1}{2\mu}\left(a^2 - x_2^2\right)\left(-\frac{dp}{dx_1}\right) \tag{5.40}$$

Thus, the velocity profile is a parabola with a maximum velocity at mid-channel as shown in Figure 5.5.

Recall that $v_1 = v_1(x_2)$. Note that $v_1 > 0$ when $\frac{dp}{dx_1} < 0$ (i.e., flow is to the right when the pressure is greater on the left than the right). The maximum velocity occurs at $x_2 = 0$. Substituting this into Eq. (5.40),

$$v_{MAX} = \left(-\frac{dp}{dx_1}\right)\frac{a^2}{2\mu} \tag{5.41}$$

The volumetric flow rate (integral of the velocity over its area of action) is

$$Q = \int_0^w \int_{-a}^a v_1 dx_2 dx_3 = \frac{2a^3w}{3\mu}\left(-\frac{dp}{dx_1}\right) \tag{5.42}$$

The average velocity is

$$v_{avg} = \frac{Q}{2aw} = \frac{a^2}{3\mu}\left(-\frac{dp}{dx_1}\right) \tag{5.43}$$

Eqs. (5.41) and (5.43) together imply

$$v_{avg} = \frac{2}{3}v_{MAX} \tag{5.44}$$

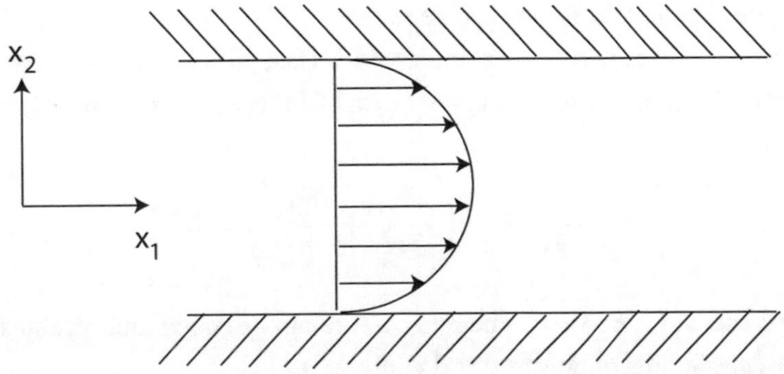

FIGURE 5.5: Velocity profile of Plane Poiseuille flow.

Finally, let us consider the consequences of $\frac{dp}{dx_1}$ being a constant. This means that $p(x_1) = C_3x_1 + C_4$. Suppose we know the inlet and outlet pressures of our system, P_{left} and P_{right}, respectively. If the channel has length L, then we can say $p(0) = P_{left}$ and $p(L) = P_{right}$. Using these conditions, we get $C_4 = P_{left}$ and $C_3 = (P_{right} - P_{left})/L$. Thus, letting $\Delta P = P_{left} - P_{right}$,

$$\frac{dp}{dx_1} = -\frac{\Delta P}{L} \tag{5.45}$$

This can be substituted into any of the Eqs. (5.39)–(5.43).

5.9.3 Extensions of Plane Poiseuille Flow

For plane Poiseuille flow, we assumed steady, fully developed flow in a rectangular channel where the dimension in the x_3 direction was much greater than the dimension in the x_2 direction. It is of interest to determine how long it takes for steady flow to develop. Consider the same rectangular channel, but v_1 now depends also upon t, i.e., $v_1 = v_1(x_2,t)$. The Navier–Stokes equations reduce to

$$-\frac{\partial p}{\partial x_1} + \mu\frac{\partial^2 v_1}{\partial x_2^2} = \rho\frac{\partial v_1}{\partial t} \tag{5.46}$$

with $v_1\ (\pm\ a,t) = 0$ and $v_1\ (x_2,\ 0) = 0$. More complicated mathematics yields the following solution [11]

$$v_1\left(x_2,t\right) = \frac{\Delta P h^2}{8\mu L}\left(1 - \frac{4x_2^2}{h^2} - \sum_{n=0}^{\infty}\frac{32\left(-1\right)^n e^{\left[\frac{-(2n+1)^2\pi^2\mu t}{\rho h^2}\right]}\cos\left[\frac{(2n+1)\pi x_2}{h}\right]}{\left(2n+1\right)^3\pi^3}\right) \tag{5.47}$$

where $h = 2a$. This expression allows us to determine how long is a reasonable time after initiation of flow to assume steady conditions (see problem 2).

Furthermore, a channel with non-infinite width (i.e., all real channels) has edge effects that cause v_1 to depend on both x_2 and x_3, i.e., $v_1 = v_1(x_2, x_3)$. Under steady conditions, the Navier–Stokes equations reduce to

$$-\frac{\partial p}{\partial x_1} + \mu\left(\frac{\partial^2 v_1}{\partial x_2^2} + \frac{\partial^2 v_1}{\partial x_3^2}\right) = 0 \tag{5.48}$$

with $v_1\ (\pm\ a, x_3) = 0$ and $v_1\ (x_2 \pm b) = 0$, where $w = 2b$. It can be shown that the steady-state solution for v_1 in a rectangular channel of height h and width w is [11]

$$v_1(x_2,x_3) = \frac{\Delta P h^2}{8\mu L}\left(1 - \frac{4x_2^2}{h^2} - \sum_{n=0}^{\infty}\frac{32(-1)^n\cosh\left[\frac{(2n+1)\pi x_3}{h}\right]\cos\left[\frac{(2n+1)\pi x_2}{h}\right]}{(2n+1)^3\pi^3\cosh\left[\frac{(2n+1)\pi w}{2h}\right]}\right) \tag{5.49}$$

from which the shear stress, flow rate, etc. can be calculated, and the magnitude of the edge effects can be determined (see problem 2).

These last two examples serve to illustrate two main points. First, they demonstrate the mathematical complexity that comes with including more details of a problem. Though the solutions to Eqs. (5.46) and (5.48) are easily obtained with elementary knowledge of partial differential equations (PDEs; see the work of Haberman [12] for details), they are much more complicated than our simple ordinary differential equation, Eq. (5.38). Second, and more importantly, they illustrate how to simplify the Navier–Stokes equations once the assumptions of the problem have been detailed.

*5.10 NON-NEWTONIAN FLUIDS

In this section, we will briefly consider some non-Newtonian fluids. Non-Newtonian fluids have nonlinear constitutive equations. Non-Newtonian fluids can be classified as inelastic or viscoelastic, depending on whether memory effects are significant. For inelastic fluids, viscosity depends on the current rate of deformation, i.e., it is not constant as it is for a Newtonian fluid. For viscoelastic fluids, the concept of memory effects means that the present state of stress depends not only on the current strain (e.g., for the elastic solid) or strain rate (e.g., the fluids discussed in this chapter), but also on aspects of the deformation history. We defer a discussion of viscoelasticity until Chapter 7.

Turning now to the constitutive equations for a class of non-Newtonian fluids, akin to Eq. (4.3), we can write the following equation describing a nonlinear viscous fluid

$$\underset{\sim}{T} = -p\underset{\sim}{I} + \alpha_1\underset{\sim}{D} + \alpha_2\underset{\sim}{D}^2 \tag{5.50}$$

where p, α_1, and α_2 are functions of temperature, density, and the scalar invariants of $\underset{\sim}{D}$. Hence, Eq. (5.50) is an isotropic tensor-valued function. Such fluids are called *Reiner–Rivlin fluids* [3]. We will consider two simplifications of this general equation.

Recall Eq. (5.15), which separated the constitutive equation of a fluid into hydrostatic pressure and viscous stress tensor components. For inelastic fluids, the viscosity coefficient in the viscous stress tensor depends on the strain-rate tensor through its invariants. For *incompressible* fluids, we know from Eq. (5.5), $I_1^D = 0$. Furthermore, there seems to be no experimental evidence that the viscosity depends on I_3^D. Hence, we have $\mu = \mu(I_2^D)$ to describe the dependence of the viscosity on the strain-rate tensor [13], where $I_2^D = \frac{1}{2}\mathrm{tr}(\underset{\sim}{D}^2)$.

The first simplification is a *power-law fluid*. The viscosity of a power-law fluid is given by

$$\mu = \mu\left(I_2^{\underset{\sim}{D}}\right) = K\left(\sqrt{I_2^{\underset{\sim}{D}}}\right)^{n-1} \tag{5.51}$$

so that

$$\underset{\sim}{T}' = 2K\left(\sqrt{I_2^{\underset{\sim}{D}}}\right)^{n-1} \underset{\sim}{D} \tag{5.52}$$

Because $I_2^{\underset{\sim}{D}}$ is a monotonically increasing function of the strain-rate, for values of $n > 1$, the apparent viscosity, $K\left(\sqrt{I_2^{\underset{\sim}{D}}}\right)^{n-1}$, increases as strain-rate increases, whereas for values of $n < 1$, the apparent viscosity decreases. Power-law fluids for which $n > 1$ are known as dilatants. A *dilatant* is a material that exhibits shear thickening (e.g., some dispersions). This is an uncommon class of materials. Power-law fluids for which $n < 1$ are known as pseudoplastics. *Pseudoplastics* are also known as shear-thinning materials (e.g., polymers).

Let us now consider another class of inelastic fluids, viscoplastic materials. Bingham plastics and Casson fluids are examples of viscoplastic materials, such as clay suspensions, drilling mud, toothpaste, and blood at low shear strain rates. A yield stress, τ_y, must be exceeded before the material will flow, though both eventually behave as Newtonian fluids. These materials behave like an elastic solid at low shear strain and then as a Newtonian fluid above a critical value. The constitutive equation for the viscous stress tensor of a *Bingham plastic* is

$$\underset{\sim}{T}' = \begin{cases} \left(\dfrac{\tau_y}{\sqrt{I_2^{\underset{\sim}{D}}}} + 2\mu\right)\underset{\sim}{D} & \frac{1}{2}\mathrm{tr}\left(\underset{\sim}{T}'^2\right) \geq \tau_y^2 \\ 0 & \frac{1}{2}\mathrm{tr}\left(\underset{\sim}{T}'^2\right) < \tau_y^2 \end{cases} \tag{5.53}$$

We defer a discussion of Casson fluids until the next chapter.

In a 1-D flow configuration, we have $\tau_{21} = \mu_{\mathrm{app}}\dot{\gamma}_1$, where τ_{21} is the shear stress in the x_2 direction on a plane of constant x_1, and $\dot{\gamma}_1$ is the shear strain-rate in the x_1 direction. Figure 5.6 compares the relationship between shear stress and strain rate in a 1-D flow configuration for various fluid models.

Finally, the viscosity of a fluid may also be time dependent. A *thixotropic fluid* is one for which μ decreases with time at constant stress (e.g., paints). A *rheopectic fluid* is one for which μ increases with time at constant stress (e.g., epoxy setting).

Demonstration. Consider the plane Couette flow of a Bingham plastic. The 1-D simplification of Eq. (5.53) is $T'_{21} = \tau_y + \mu\dfrac{\partial v_1}{\partial x_2}$ (see problem 3). A shear stress is applied to the upper plate. What is the velocity of the upper plate as a function of the applied shear stress?

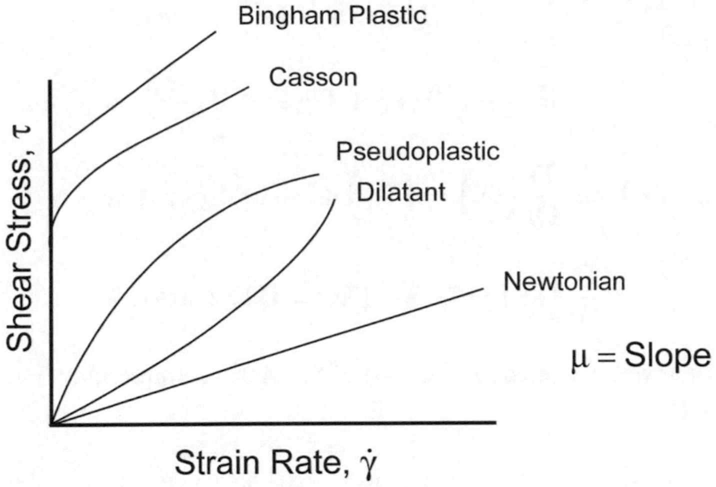

FIGURE 5.6: Comparison of shear stress versus strain rate for a Newtonian and several non-Newtonian fluids.

Solution. The Bingham plastic does not flow until the yield stress is overcome. Hence, for $T'_{21} < \tau_y$, the velocity of the upper plate is zero. For $T'_{21} \geq \tau_y$, we solve the equation $T'_{21} = \tau_y + \mu \dfrac{\partial v_1}{\partial x_2}$, subject to the boundary condition $v_1(x_2 = 0) = 0$. The solution is $v_1(x_2) = \dfrac{(T'_{21} - \tau_y)\, x_2}{\mu}$. Thus, the velocity of the upper plate is $v_1(d) = \dfrac{(T'_{21} - \tau_y)\, d}{\mu}$.

*5.11 VORTICITY VECTOR

Recall that the velocity gradient, $\nabla \vec{v}$, can be written as (Eq. (2.35))

$$\nabla \vec{v} = \underset{\sim}{D} + \underset{\sim}{W} \tag{5.54}$$

where

$$\underset{\sim}{D} = \frac{1}{2}\left[\left(\nabla \vec{v}\right) + \left(\nabla \vec{v}\right)^T\right] \tag{5.55}$$

and

$$\underset{\sim}{W} = \frac{1}{2}\left[\left(\nabla \vec{v}\right) - \left(\nabla \vec{v}\right)^T\right]$$

Because $\underset{\sim}{W}$ is antisymmetric, its diagonal elements are 0, and the only three independent off-diagonal elements are W_{12}, W_{31}, and W_{23}. Thus, the spin tensor is equivalent to a vector $\vec{\omega}$ expressed by

$$\underset{\sim}{W}\vec{x} = \vec{\omega} \times \vec{x} \tag{5.56}$$

$$\vec{\omega} = -\left(W_{23}\vec{e}_1 + W_{31}\vec{e}_2 + W_{12}\vec{e}_3\right) \tag{5.57}$$

Recall Eq. (2.33). From $\frac{\mathrm{D}}{\mathrm{D}t}\left(\mathrm{d}\vec{x}\right) = \left(\nabla\vec{v}\right)\mathrm{d}\vec{x}$ and Eq. (5.54) we get

$$\frac{\mathrm{D}}{\mathrm{D}t}\left(\mathrm{d}\vec{x}\right) = \underset{\sim}{D}\mathrm{d}\vec{x} + \underset{\sim}{W}\mathrm{d}\vec{x} = \underset{\sim}{D}\mathrm{d}\vec{x} + \vec{\omega} \times \mathrm{d}\vec{x} \tag{5.58}$$

Thus, $\vec{\omega}$ is the angular velocity vector of that part of the motion representing the rigid body rotation of a material particle.

$$\vec{\omega} = \frac{1}{2}\left(\frac{\partial v_3}{\partial x_2} - \frac{\partial v_2}{\partial x_3}\right)\vec{e}_1 + \frac{1}{2}\left(\frac{\partial v_1}{\partial x_3} - \frac{\partial v_3}{\partial x_1}\right)\vec{e}_2 + \frac{1}{2}\left(\frac{\partial v_2}{\partial x_1} - \frac{\partial v_1}{\partial x_2}\right)\vec{e}_3 \tag{5.59}$$

Let $\vec{\zeta} = 2\,\vec{\omega}$ be the vorticity vector. Then,

$$\vec{\zeta} = 2\vec{\omega} = \left(\frac{\partial v_3}{\partial x_2} - \frac{\partial v_2}{\partial x_3}\right)\vec{e}_1 + \left(\frac{\partial v_1}{\partial x_3} - \frac{\partial v_3}{\partial x_1}\right)\vec{e}_2 + \left(\frac{\partial v_2}{\partial x_1} - \frac{\partial v_1}{\partial x_2}\right)\vec{e}_3 \tag{5.60}$$

and $2\underset{\sim}{W}$ is the vorticity tensor.

Demonstration. Find the vorticity vector for the simple shearing flow $v_1 = kx_2$, $v_2 = v_3 = 0$.

Solution.

$\zeta_1 = \left(\frac{\partial v_3}{\partial x_2} - \frac{\partial v_2}{\partial x_3}\right) = 0$, $\zeta_2 = 0$, $\zeta_3 = -k$. So, $\vec{\zeta} = -k\vec{e}_3$. The angular velocity vector, $\frac{1}{2}\vec{\zeta}$, is normal to the x_2 - x_3 plane. The minus sign means the spinning is clockwise looking from the positive side of x_3.

*5.12 IRROTATIONAL FLOW

If the vorticity vector, $\vec{\zeta}$, or the vorticity tensor, $2\underset{\sim}{W}$, corresponding to a velocity field is zero in some region of the fluid for some time interval, the flow is called irrotational. Let $\phi(x_1, x_2, x_3, t)$ be a scalar function, and let the velocity components be derived from ϕ by

$$v_1 = -\frac{\partial\phi}{\partial x_1} \quad v_2 = -\frac{\partial\phi}{\partial x_2} \quad v_3 = -\frac{\partial\phi}{\partial x_3} \tag{5.61}$$

or

$$v_i = -\frac{\partial \phi}{\partial x_i} \tag{5.62}$$

Then by Eq. (5.60), the vorticity vector becomes

$$\zeta_1 = \left(\frac{\partial v_3}{\partial x_2} - \frac{\partial v_2}{\partial x_3}\right) = -\frac{\partial^2 \phi}{\partial x_2 \partial x_3} + \frac{\partial^2 \phi}{\partial x_3 \partial x_2} = 0, \quad \zeta_2 = 0, \quad \zeta_3 = 0 \tag{5.63}$$

Thus, a scalar function $\phi(x_1, x_2, x_3, t)$ defines an irrotational flow field if $v_i = -\frac{\partial \phi}{\partial x_i}$. Of course, arbitrary functions ϕ work if, in addition to Eq. (5.62), they also satisfy the continuity equation.

Recall that for an incompressible fluid the continuity equation is

$$\frac{\partial v_i}{\partial x_i} = 0 \tag{5.64}$$

Eq. (5.62) into Eq. (5.64) yields

$$\frac{\partial^2 \phi}{\partial x_i^2} = 0 \tag{5.65}$$

Eq. (5.65) is the Laplace equation for ϕ, i.e.,

$$\nabla^2 \phi = \frac{\partial^2 \phi}{\partial x_1^2} + \frac{\partial^2 \phi}{\partial x_2^2} + \frac{\partial^2 \phi}{\partial x_3^2} = 0 \tag{5.66}$$

So, for irrotational flow, there exists a scalar, ϕ, which satisfies $v_i = -\frac{\partial \phi}{\partial x_i}$ and $\nabla^2 \phi = 0$.

5.12.1 Irrotational Flow of an Inviscid Incompressible Fluid

An *inviscid fluid* is defined by

$$T_{ij} = -p\delta_{ij} \tag{5.67}$$

The equations of motion for an inviscid fluid are

$$\rho\left(\frac{\partial v_i}{\partial t} + v_j \frac{\partial v_i}{\partial x_j}\right) = \rho B_i - \frac{\partial p}{\partial x_i} \tag{5.68}$$

or, in invariant form,

$$\rho \frac{\mathrm{D}\vec{v}}{\mathrm{D}t} = -\nabla p + \rho \vec{B} \tag{5.69}$$

Eq. (5.68) is known as *Euler's equation of motion*. For an incompressible fluid with homogeneous density, irrotational flows are always possible provided that the body forces acting are derivable from a potential Ω, i.e.,

$$B_i = -\frac{\partial \Omega}{\partial x_i} \tag{5.70}$$

For example, in the case of gravity with positive x_3 pointing up,

$$\Omega = gx_3 \tag{5.71}$$

yielding $B_1 = B_2 = 0$ and $B_3 = -g$. Eq. (5.70) into Eq. (5.68) results in

$$\frac{\partial v_i}{\partial t} + v_j \frac{\partial v_i}{\partial x_j} = -\frac{\partial}{\partial x_i}\left(\frac{p}{\rho} + \Omega\right) \tag{5.72}$$

But, for irrotational flow, $\frac{\partial v_i}{\partial x_j} = \frac{\partial v_j}{\partial x_i}$ (see problem 5), so that $v_j \frac{\partial v_i}{\partial x_j} = v_j \frac{\partial v_j}{\partial x_i} = \frac{1}{2}\frac{\partial}{\partial x_i}\left(v_j^2\right)$.

Note that $v_j^2 = v_j v_j$, which is a summation. Hence, $v_j \frac{\partial v_i}{\partial x_j} = \frac{1}{2}\frac{\partial v^2}{\partial x_i}$, where $v^2 = v_1^2 + v_2^2 + v_3^2$ is the square of the magnitude of the velocity (i.e., speed squared). Eq. (5.72) becomes $\frac{\partial v_i}{\partial t} + \frac{\partial}{\partial x_i}\left(\frac{v^2}{2} + \frac{p}{\rho} + \Omega\right)$. But, $v_i = -\frac{\partial \phi}{\partial x_i}$ from Eq. (5.62), so

$$\frac{\partial}{\partial x_i}\left(-\frac{\partial \phi}{\partial t} + \frac{v^2}{2} + \frac{p}{\rho} + \Omega\right) = 0 \tag{5.73}$$

or

$$-\frac{\partial \phi}{\partial t} + \frac{v^2}{2} + \frac{p}{\rho} + \Omega = f(t) \tag{5.74}$$

where $f(t)$ is some arbitrary function of t. If the flow is steady, i.e., at every fixed point nothing changes with time, then

$$\frac{v^2}{2} + \frac{p}{\rho} + \Omega = C \tag{5.75}$$

where C is a constant. Eq. (5.75) is *Bernoulli's equation*. Eq. (5.74) is the general form of Bernoulli's equation. This equation applies when the effect of viscosity can be neglected. For whatever function ϕ, as long as $v_i = -\dfrac{\partial \phi}{\partial x_i}$ and $\nabla^2 \phi = 0$, the equations of motion can be integrated to give Bernoulli's equation! Let us briefly consider the assumptions leading to Eq. (5.75). We have assumed 1. incompressible flow of an 2. inviscid fluid that is 3. steady 4. along a streamline subject only to a 5. body force derived from a potential.

Demonstration. Given $\phi = x_1^3 - 3x_1x_2^2$,

(a) Show that ϕ satisfies the Laplace equation.

(b) Find the irrotational velocity field.

(c) Find the pressure distribution for an incompressible, inviscid, homogeneous fluid if $p(0,0,0) = p_0$ and $\Omega = gx_3$.

Solution.

(a) $\dfrac{\partial^2 \phi}{\partial x_1^2} = 6x,\ \dfrac{\partial^2 \phi}{\partial x_2^2} = -6x,\ \dfrac{\partial^2 \phi}{\partial x_3^2} = 0$, so that $\nabla^2 \phi = \dfrac{\partial^2 \phi}{\partial x_1^2} + \dfrac{\partial^2 \phi}{\partial x_2^2} + \dfrac{\partial^2 \phi}{\partial x_3^2} =$ $6x + -6x + 0 = 0$

(b) From $v_i = -\dfrac{\partial \phi}{\partial x_i}$,

$$v_1 = -\frac{\partial \phi}{\partial x_1} = -3x_1^2 + 3x_2^2$$

$$v_2 = -\frac{\partial \phi}{\partial x_2} = 6x_1x_2$$

$$v_3 = -\frac{\partial \phi}{\partial x_3} = 0$$

(c) At (0,0,0), we have $v_1 = v_2 = v_3 = 0$, $p = p_0$, and $\Omega = gx_3 = 0$.

From Bernoulli's equation, $\dfrac{v^2}{2} + \dfrac{p}{\rho} + \Omega = C$, which means $C = \dfrac{p_0}{\rho}$. At other points, Bernoulli's equation yields

$$p = \rho C - \frac{\rho v^2}{2} - \rho\Omega$$

$$p = p_0 - \frac{\rho}{2}\left(v_1^2 + v_2^2\right) - \rho g x_3$$

$$p = p_0 - \frac{\rho}{2}\left[9\left(x_2^2 - x_1^2\right)^2 + 36x_1^2x_2^2\right] - \rho g x_3$$

5.13 PROBLEMS

1. Derive the Navier–Stokes equations, Eq. (5.26), from Cauchy's equation of motion and the constitutive equation for an incompressible Newtonian fluid. Note that these equations apply only to an incompressible Newtonian fluid.

Next, consider gravity as the only acting body force (i.e., let the x_3 axis point vertically down such that $B_1 = B_2 = 0$, and $B_3 = g$) and the constitutive equation for the fluid to be Eq. (5.15). Show that the equation of motion in the x_3 direction is

$$\rho\left(\frac{\partial v_3}{\partial t} + v_i\frac{\partial v_3}{\partial x_i}\right) = \rho g - \frac{\partial p}{\partial x_3} + \frac{\partial T'_{i3}}{\partial x_i}$$

2. In this problem, we will look at various aspects of the equations for the extensions of plane Poiseuille flow.
   (a) Show how Eqs. (5.46) and (5.48) are obtained from the Navier–Stokes equations.
   (b) Show in the limit as $t \to \infty$ that Eq. (5.47) reduces to Eq. (5.40) with $a = h/2$. Show also that, as $w \to \infty$, Eq. (5.49) reduces to Eq. (5.40).
   (c) Consider Eq. (5.46). In terms of h, μ, and ρ, how long does it take for the velocity at mid-channel to be within 5% of its final value? Within 1%?

(Hint: First calculate $\dfrac{v_1(x_2,t)}{v_1^{\text{s.s.}}(x_2)}$, where $v_1^{\text{s.s.}}(x_2)$ is the steady-state solution solved for in the text. Then, use Mathematica©'s 'N[ ]' command to evaluate the sum.)

   (d) Consider Eq. (5.49). Find the shear stress at the wall ($x_2 = -h/2$). What aspect ratio, i.e., how much greater does w have to be than h, yields an average wall shear stress within 5% of the solution when we considered the width to be infinite? Within 1%?

(Hint: Calculate $\dfrac{\text{avg}\,T_{12}(x_2 = -h/2)}{T_{12}^{\infty}(x_2 = -h/2)}$, where $T_{12}^{\infty}(x_2 = -h/2)$ is the wall shear stress for a channel of infinite width. Note that $\text{avg}T_{12}(x_2 = -h/2) = \dfrac{1}{w}\displaystyle\int_{-w/2}^{w/2} T_{12}(x_2 = -h/2, x_3)\mathrm{d}x_3$.)

3. Consider a flow described by $v_1 = v_1(x_2)$. Simplify Eqs. (5.52) and (5.53) to find T'_{12} for a power-law fluid and for a Bingham plastic. In the case of a power-law fluid, what values of K and n yield the constitutive equation for a Newtonian fluid? In the case of a Bingham plastic, what value of τ_y results in a Newtonian fluid?
4. Consider plane Poiseuille flow of a thixotropic, incompressible Newtonian fluid. For a thixotropic fluid, μ decreases with time. Assuming μ depends on time as $\mu(t) = 10e^{-3t} + 2$,

plot the velocity at mid-channel from $t = 0$ to $t = 5$. What is the effect of μ on wall shear stress?

5. Irrotational flow is where fluid elements in the flow field do not undergo rotation. Rotational flow is described by the vorticity vector, $\vec{\zeta}$, which is written as $\vec{\zeta} = -2(W_{23}\vec{e}_1 + W_{31}\vec{e}_2 + W_{12}\vec{e}_3) = -\varepsilon_{ijk} W_{jk}\vec{e}_i$, where W_{ij}'s are components of the antisymmetric spin tensor, and ε_{ijk} is the permutation symbol. For irrotational flow, $\vec{\zeta} = 0$. Show that this implies $\dfrac{\partial v_i}{\partial x_j} = \dfrac{\partial v_j}{\partial x_i}$.
6. Simplify the Navier–Stokes equations by successively making the following assumptions:
   (a) Steady-state flow (i.e., all properties independent of time)
   (b) Irrotational flow (i.e., $\partial v_i/\partial x_j = \partial v_j/\partial x_i$)
   (c) Incompressibility (i.e., $\partial v_i/\partial x_i = 0$)
   (d) When the only body force is gravity (i.e., $\vec{B} = -g\vec{e}_3 = -g\nabla x_3$)

   Now, integrate the expression. This result is a powerful and widely used equation, known as Bernoulli's equation, which relates pressure changes to changes in elevation and velocity.
7. Given the following flow field of an incompressible Newtonian fluid

$$v_1 = kx_1 \quad v_2 = -3kx_2 \quad v_3 = cx_3$$

   with k, c constants.
   (a) Determine the relationship between c and k.
   (b) Find the stress tensor.
   (c) Neglecting body forces, derive the pressure distribution, with $p = p_0$ at the origin, directly from the Navier–Stokes equations.
   (d) Find the acceleration field in terms of the pressure field.

8) A velocity profile for blood plasma, a viscous fluid, is as follows:

$$v_1 = 4x_1^2 x_2 - 9x_2^2$$
$$v_2 = 12x_3^3 - 6x_1 x_2^2$$
$$v_3 = 4x_1 x_2 x_3$$

   (a) Is this velocity profile consistent with the assumption of an incompressible fluid?
   (b) What is the simplest linear constitutive equation you would use in this situation that relates stress to strain rate?

9) Over spring break, you decided to scale Mt. Everest. While regaling your friends with stories from your adventure, one asks how you were able to handle the pressure change occurring from base camp to the summit. Being a fellow engineer, your friend wants precise information regarding the pressure distribution you encountered while climbing to the World's highest point. The equation for fluid hydrostatics is $\nabla p = \rho \vec{B}$. Consider air to be an ideal gas, which, with appropriate scaling, has the equation of state $p = \rho RT$. *We wish to calculate the pressure as a function of distance from base camp in the column of air surrounding Mt. Everest.* In this column of air, the temperature can be written as a linear function of x_3, $T = T_{BC} - \alpha x_3$ (where T_{BC} is the cold base camp temperature). Furthermore, the only force acting on this column of air is gravity (i.e., $\vec{B} = -g\vec{e}_3$ with positive x_3 directed upward). The pressure at base camp (i.e., $x_3 = 0$) is P_{BC}. You may assume that the acceleration due to gravity did not change appreciably during your climb.

• • • •

CHAPTER 6

Blood and Circulation

6.1 INTRODUCTION

Blood is the main fluid of the circulatory system, functioning in transport and helping to maintain homeostasis. In terms of transport, blood delivers oxygen picked up in the lungs and nutrients from internal stores (e.g., proteins and carbohydrates) to cells throughout the body and then carries away waste (e.g., lactic acid) to other organs for breakdown and carbon dioxide back to the lungs for exhalation. However, blood has several other important roles. Blood is involved in controlling the body's pH by acting as an extensive buffer, serves to regulate temperature by transporting heat from the body's core to the exterior, and is the staging ground for many cells of the immune system.

In this chapter, our goal is to study the rheological behavior of blood to shed insight into its bulk transport capacity. To accomplish this task, we will develop governing equations for blood flow, including constitutive equations and expressions for flow rate. However, we must start with some basic properties and concepts. The forces driving blood flow are gravity, pressure gradients, and muscle contraction (veins), while the forces opposing blood flow are shear forces due to viscosity and turbulence. The pressure at a given point is the sum of static pressure due to gravity, pressure due to beating of the heart, and frictional loss in vessels.

The flow of blood has often been described in the c-g-s system instead of the m-kg-s system. Hence, some useful unit conversions when talking about blood flow are the following:

$$\text{Viscosity} \rightarrow 100\ \text{cP} = 0.1\ \text{kg/m·s} = 0.1\ \text{N·s/m}^2 = 0.1\ \text{Pa·s} = 1\ \text{dyne·s/cm}^2$$

$$\text{Force} \rightarrow 1\ \text{N} = 10^5\ \text{dyne}$$

6.2 BASICS AND MATERIAL PROPERTIES OF BLOOD

Blood is a viscous fluid mixture of plasma and cells. The blood volume is ~5 L, of which 3 L are plasma and 2 L are cells. The hematocrit is defined as the fraction of blood volume due to cells, mainly red blood cells (RBCs), and is approximately 40%. RBCs, also known as erythrocytes, account for more than 95% of blood's cellular composition. The remainder is made up of platelets

(~5%) and leukocytes (~0.1%), or white blood cells (WBCs). The effects of platelets and WBCs are negligible on macroscopic flow. If the cellular components of blood are removed, one is left with plasma, which is 90% H_2O, ~7–8% protein, and ~2% various organic and inorganic substances. One obtains serum if the plasma is allowed to clot; hence, serum is essentially the same as plasma, except clotting proteins have been removed.

Whole blood is considered to exhibit a viscosity around $\mu = 3$ to 6 cP. This low viscosity value is at high shear rates and 37°C. Under different conditions, whole blood viscosity can approach 100 cP. The density of whole blood is $\rho = 1.056$ g/cm^3. If we look only at plasma, $\mu = 1.2$ cP and $\rho = 1.024$ g/cm^3. Red blood cells exhibit $\rho = 1.098$ g/cm^3. Their diameter is ~7.65 ± 0.6 μm, and their thickness is ~1–2.7 μm. Note, however, that capillaries are ~4–10 μm in diameter, so RBCs must deform and flow in single file.

6.3 REYNOLDS NUMBERS FOR BLOOD

In Chapter 5, we learned that, in general, when Re > 2,300, turbulent flow ensues. Within the human circulatory system, typical values of Re are as follows:

Human aorta → Re > 3,000
Small arteries and veins → Re ~ 1
Microvessels →Re < 1
Capillaries → Re ~ 10^{-2}

Therefore, flow is laminar in vessels that are sufficiently small. Turbulence dissipates energy, resulting from a higher friction coefficient, and has been strongly implicated in atherogenesis, which most often affects larger vessels.

6.4 NON-NEWTONIAN BEHAVIOR OF BLOOD

Recall Eq. (5.16) for Newtonian fluids, $T'_{ij} = \lambda D_{kk}\delta_{ij} + 2\mu D_{ij}$, where the viscosity, μ, is constant. Consider the simple 1-D case, $T_{12} = 2\mu D_{12}$, or, using an older notation,

$$\tau = \mu\dot{\gamma} \tag{6.1}$$

In Eq. (6.1), τ is the shear stress, and $\dot{\gamma}$ is the strain rate (the dot over γ indicating a time rate of change). Figure 6.1 shows shear stress as a function of strain rate for blood, from which we can see

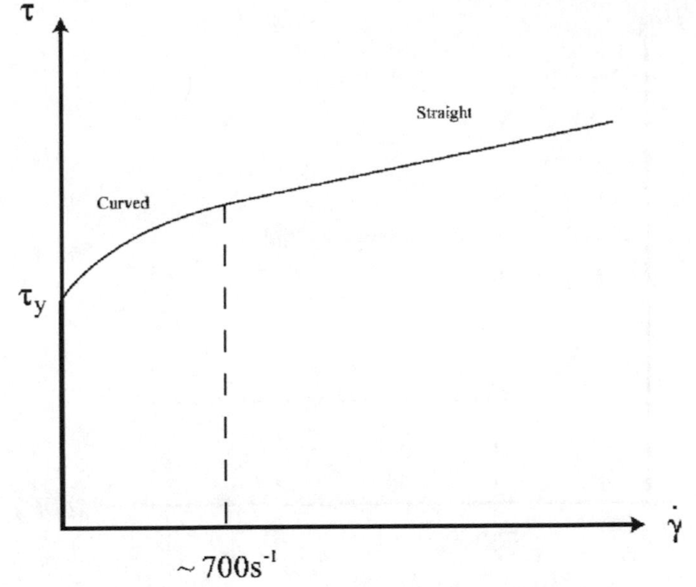

FIGURE 6.1: Shear stress versus strain rate for blood. The slope of the curve is not constant, indicating that the viscosity of blood is not constant.

that the relationship is clearly nonlinear, i.e., μ is not constant. However, at high strain rates, the curve is essentially linear, representing a Newtonian fluid.

The transition of blood to Newtonian behavior (constant μ) is a function of hematocrit and occurs at about $\dot{\gamma} > 700\ s^{-1}$ for normal values [14]. Fibrinogen, and its effect on facilitating RBC adhesion, is largely responsible for the non-Newtonian behavior of blood. At low strain rates, RBCs stack into structures called *rouleaux*, which can clump into larger aggregates. Aggregation of RBCs is strongly dependent on shear strain rate, as they break up as strain rate increases. Thus, with increasing strain rate, the viscosity of blood decreases. The end result is the curved part of the τ vs. $\dot{\gamma}$ plot. This portion of the curve resembles a pseudoplastic for which as

$$\uparrow \dot{\gamma} \Rightarrow \downarrow \mu$$

Furthermore, blood, similar to a Bingham plastic, has a yield stress, τ_y. The yield stress of blood is ~0.05 dyne/cm^2 [15]. This behavior of blood is represented by the very beginning of the curve.

Based on these empirical observations, *blood is classified as a thixotropic Casson fluid*, although we will neglect time dependence in our analysis. This is the simplest constitutive equation for blood, though others have been put forth. Figure 6.2 shows the derivative of the τ vs. $\dot{\gamma}$ curve, yielding the viscosity as a function of shear strain rate.

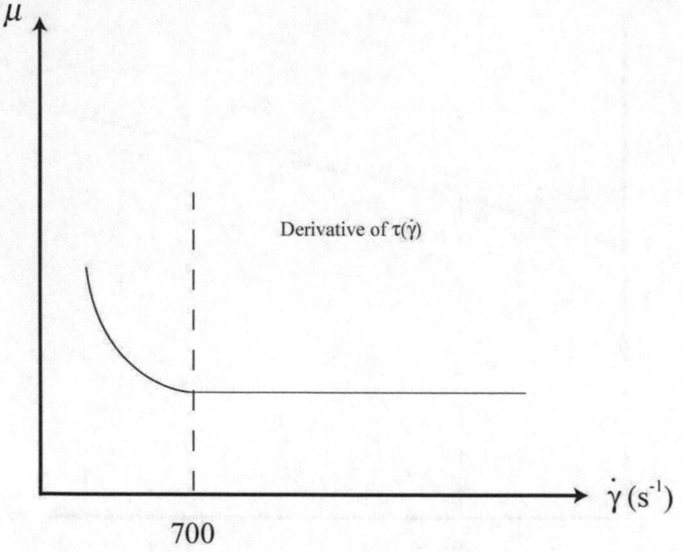

FIGURE 6.2: Derivative of the τ versus $\dot{\gamma}$ curve for blood.

6.5 CASSON EQUATION

Blood seems to obey the following empirically derived equation

$$\sqrt{\tau} = \sqrt{\tau_y} + \sqrt{\eta\dot{\gamma}} \tag{6.2}$$

where τ is the shear stress, τ_y is the yield stress, η a constant, and $\dot{\gamma}$ the shear strain rate. For blood, $\eta \approx 4$ cP (and depends on the hematocrit), and $\tau_y \approx 0.05$ dyne/cm^2 (which is very small). One can use the Casson equation for hematocrits $\leq 40\%$. Substituting Eq. (6.1) into Eq. (6.2) yields

$$\mu\left(\dot{\gamma};\tau_y,\eta\right) = \frac{\left(\sqrt{\tau_y} + \sqrt{\eta\,|\dot{\gamma}|}\right)^2}{|\dot{\gamma}|} \tag{6.3}$$

From Eq. (6.3), we note that as $\dot{\gamma}$ becomes large, $\mu \to \eta$.

6.6 BLOOD RHEOLOGY

If the shear strain rate is sufficiently high, blood behaves as an isotropic, incompressible Newtonian fluid. The isotropic assumption here means that, when τ and $\dot{\gamma} = 0$, blood cells have no preferred orientation. Furthermore, blood is incompressible at physiological pressures. Recall the equation for an incompressible Newtonian flow, Eq. (5.23)

$$T_{ij} = -p\delta_{ij} + 2\mu D_{ij} \tag{6.4}$$

How can we generalize this Newtonian equation to account for blood's non-Newtonian behavior? We will introduce a "positively valued" second invariant, J_2. Recall Eqs. (1.78)–(1.81), which say that a symmetric tensor has three invariants

$$\begin{aligned} I_1 &= T_{ii} \\ I_2 &= \frac{1}{2}(T_{ii}T_{jj} - T_{ij}T_{ji}) \\ I_3 &= \det(\underset{\sim}{T}) \end{aligned} \tag{6.5}$$

Let

$$J_2 = \frac{1}{2}I_1^2 - I_2 = \frac{1}{2}T_{ij}T_{ji} \tag{6.6}$$

Because $I_1 = 0$ for incompressible fluids, J_2 is essentially a sign change for I_2. For the strain-rate tensor,

$$J_2(\underset{\sim}{D}) = \frac{1}{2}D_{ij}D_{ij} \tag{6.7}$$

where we have switched the order of the last indices, as $\underset{\sim}{D}$ is symmetric. For non-Newtonian behavior, the viscosity can be expressed as a function of J_2 such that

$$T_{ij} = -p\delta_{ij} + 2[\mu(J_2)]D_{ij} \tag{6.8}$$

For a simple 1-D flow,

$$\dot{\gamma} = 2D_{12} \tag{6.9}$$

and

$$J_2 = \frac{1}{2}\left(D_{12}^2 + D_{21}^2\right) = D_{12}^2 \tag{6.10}$$

Substituting Eq. (6.10) into (6.9),

$$\dot{\gamma} = 2\sqrt{J_2} \tag{6.11}$$

In general, Eq. (6.11) into Eq. (6.3) plus some algebra yields

$$\mu(J_2) = \frac{\left[\left(\frac{\tau_y}{2}\right)^{1/2} + \left(\eta^2 J_2\right)^{1/4}\right]^2}{\sqrt{J_2}} \tag{6.12}$$

where J_2 is given by Eq. (6.7).

6.7 SUMMARY

Blood viscosity can exhibit three behaviors based on three flow regimes.

1. *Elastic.* Before the yield stress is overcome, blood behaves like an elastic solid, e.g., a linearly elastic solid may be proposed. Below the yield stress, blood does not flow and shear strain rate is zero. However, at some point, the yield stress will be overcome. We can describe the yield condition using $J_2(\underset{\sim}{T}^0)$, as is often done in the theory of plasticity [14], where $\underset{\sim}{T}^0$ is the deviatoric part of the stress tensor. Recall Chapter 3, problem 1,

$$T^0_{ij} = T_{ij} - \frac{1}{3} T_{kk}\delta_{ij} \tag{6.13}$$

The yield condition is given by

$$J_2(\underset{\sim}{T}^0) = \frac{1}{2} T^0_{ij} T^0_{ij} \tag{6.14}$$

For $J_2(\underset{\sim}{T}^0) < K$, blood remains elastic (obeys Hooke's law, and $D_{ij} = 0$), while for $J_2(\underset{\sim}{T}^0) \geq K$, blood yields and flows. $K \sim 4\text{–}25 \times 10^{-6}$ N^2/m^4 and depends on the hematocrit. The constitutive equation is (recalling Eqs. (4.30) and (4.38) applied to an incompressible material)

$$T_{ij} = \frac{2E_Y}{3} E_{ij} \tag{6.15}$$

2. *Non-Newtonian flow.* Under this regime, blood yields and flows. Given $J_2(\underset{\sim}{T}^0) \geq K$, the constitutive equation is given by Eqs. (6.7), (6.8), and (6.12)

$$T_{ij} = -p\delta_{ij} + 2\left\{\frac{\left[\left(\frac{\tau_y}{2}\right)^{1/2} + \left(\eta^2 \frac{1}{2} D_{kl} D_{kl}\right)^{1/4}\right]^2}{\sqrt{\left(\frac{1}{2} D_{kl} D_{kl}\right)}}\right\} D_{ij} \tag{6.16}$$

when $J_2(\underset{\sim}{D}) < c$. From Eq. (6.11), and a transition around $\dot{\gamma} \sim 700$ s^{-1}, $c = 122{,}500$ s^{-2}.

TABLE 6.1: Summary of the constitutive behavior of blood

REGIME	CRITERIA	CONSTITUTIVE EQUATION
Elastic	$J_2(\underset{\sim}{T}^0) < K$	$T_{ij} = \frac{2E_Y}{3}E_{ij}$
Non-Newtonian fluid	$J_2(\underset{\sim}{T}^0) \geq K$ $J_2(\underset{\sim}{D}) < c$	$T_{ij} = -p\delta_{ij} + 2[\mu(J_2(\underset{\sim}{D}))]\, D_{ij}$
Newtonian fluid	$J_2(\underset{\sim}{T}^0) \geq K$ $J_2(\underset{\sim}{D}) \geq c$	$T_{ij} = -p\delta_{ij} + 2\mu D_{ij}$

$K \sim 4\text{–}25 \times 10^{-6}$ N^2/m^4 and $c \sim 122{,}500$ s^{-2}.

3. *Newtonian flow.* At high shear strain rates, past the transition point of $\sim \dot{\gamma} > 700$ s^{-1} with $J_2(\underset{\sim}{T}^0) \geq K$ and $J_2(\underset{\sim}{D}) \geq c$, the constitutive equation is given by Eq. (5.23)

$$T_{ij} = -p\delta_{ij} + 2\mu D_{ij} \tag{6.17}$$

where $\mu \approx 4$ cP. Table 6.1 summarizes the constitutive behavior of blood. We will now use these concepts to examine the flow of blood in a tube, e.g., an artery.

Though, perhaps, seemingly overcomplicated, the development of Eq. (6.16) is an essential step in describing blood rheology. The Casson equation is sufficient for problems in which the flow is simple, but complicated flows require this more robust constitutive equation for their analyses.

6.8 LAMINAR FLOW OF BLOOD IN A TUBE

Consider the flow of blood in a circular cylindrical tube of diameter d (e.g., an artery). We will make the following assumptions: 1. laminar, 2. steady, 3. axisymmetric, and 4. fully developed flow (no entrance effects). Considering assumption 4, for fully developed flow in a cylindrical tube the entrance length is given by

$$L_e = (0.06)\text{Re } d \tag{6.18}$$

where Re is the Reynolds number and d is the tube diameter. L_e describes the minimum distance of flow upon entrance into the tube for which variations of velocity in the direction coincident with the tube's long axis can be neglected. Note that the distance between successive artery bifurcations may not satisfy assumption 4. Furthermore, for laminar flow in cylindrical tubes, Re < 2,100 [11].

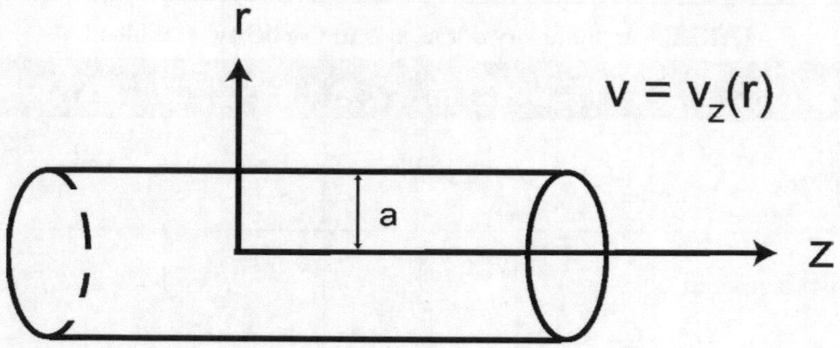

FIGURE 6.3: Geometry for Poiseuille flow in a cylindrical tube.

In the derivations below we will use cylindrical coordinates, r, θ, and z (see Figure 6.3). Assuming axisymmetric flow, all variables are independent of θ. Furthermore, past the entrance length, v_z can no longer depend on z. Hence, $v_z = v_z(r)$.

Newtonian flow regime. We have cylindrical Poiseuille flow (see Section 5.9), where Poiseuille flow is laminar, steady, pressure-driven flow of an incompressible Newtonian fluid. In cylindrical coordinates (∇ has different expressions depending on the coordinate system used) the Navier–Stokes equations give for 1-D flow, analogous to Eq. (5.38),

$$\frac{\mathrm{d}p}{\mathrm{d}z} = \frac{\mu}{r}\frac{\mathrm{d}}{\mathrm{d}r}\left(r\frac{\mathrm{d}v_z}{\mathrm{d}r}\right) \tag{6.19}$$

The "no-slip" boundary condition is

$$v_z\,(r = a) = 0 \tag{6.20}$$

The other boundary condition is a symmetry condition describing the fact that v_z is maximum at the center of the tube ($r = 0$). Mathematically,

$$\left.\frac{\mathrm{d}v_z}{\mathrm{d}r}\right|_{r=0} = 0 \tag{6.21}$$

From Eqs. (6.19)–(6.21), it can be shown (see problem 1)

$$v_z\,(r) = \frac{1}{4\mu}\left(a^2 - r^2\right)\left(-\frac{\mathrm{d}p}{\mathrm{d}z}\right) \tag{6.22}$$

$$Q = \int_A v_z \mathrm{d}A = 2\pi\int_0^a v_z r \mathrm{d}r = \frac{\pi a^4}{8\mu}\left(-\frac{\mathrm{d}p}{\mathrm{d}z}\right) \tag{6.23}$$

$$v_{MAX} = \frac{a^2}{4\mu}\left(-\frac{dp}{dz}\right) \tag{6.24}$$

$$v_{avg} = \frac{a^2}{8\mu}\left(-\frac{dp}{dz}\right) = \frac{1}{2}v_{MAX} \tag{6.25}$$

Note the similarity of Eqs. (6.22)–(6.25) with Eqs. (5.40)–(5.44). This example illustrates how the geometry of the problem can affect the solution.

Casson (non-Newtonian) flow regime. Consider the force balance on a fluid element (see Figure 6.4).

$$\tau(2\pi r)L + P_2\left(\pi r^2\right) = P_1(\pi r^2) \Rightarrow 2\tau = \frac{-(P_2 - P_1)}{L}r$$

As $L \to 0$,

$$\tau = -\frac{r}{2}\frac{dp}{dz} \tag{6.26}$$

Note that $\tau|_{r=0} = 0 \Rightarrow \tau|_{r=0} = 0 < \tau_y$! The plot of the shear stress profile is shown in Figure 6.5, where the slope of the line is $-\frac{1}{2}\frac{dp}{dz}$.

From Eq. (6.26), the shear stress at the wall of the cylinder is

$$\tau_w = \frac{a}{2}\left(-\frac{dp}{dz}\right) \tag{6.27}$$

At some radial distance from the tube's center, the shear stress will overcome the yield stress. Let this distance be r_c. At r_c the shear stress is equal to the yield stress and is given by

$$\tau_y = \frac{r_c}{2}\left(-\frac{dp}{dz}\right) \tag{6.28}$$

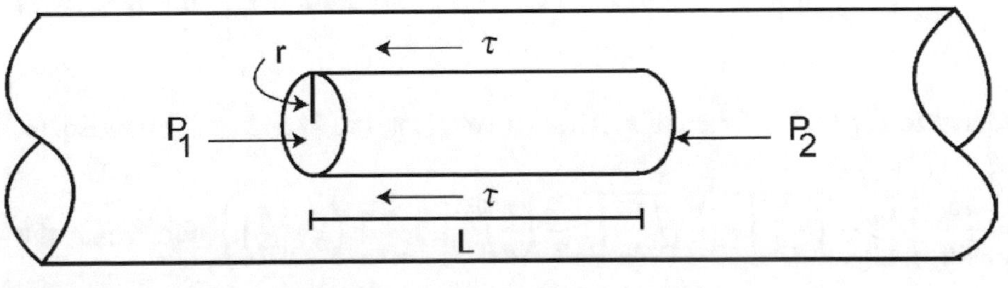

FIGURE 6.4: Force balance on a fluid element in cylindrical coordinates.

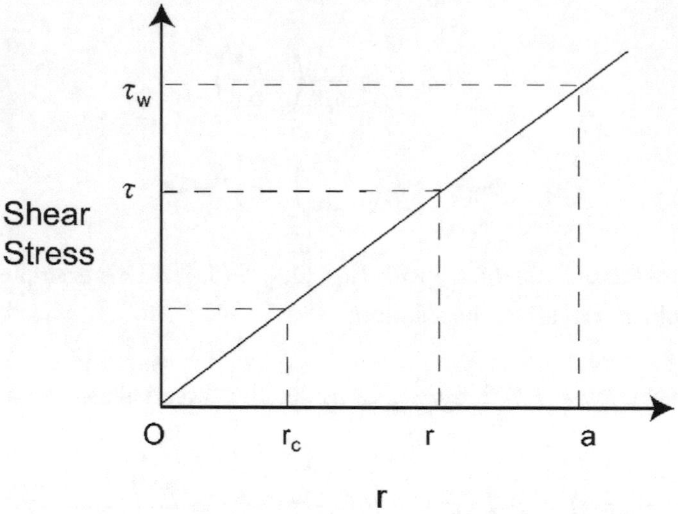

FIGURE 6.5: Plot of shear stress as a function of distance from the center tube.

Recall the notation $\tau = \mu\dot{\gamma}$. With the positive z-axis to the right in Figure 6.4, $-\tau = \mu \frac{dv_z}{dr}$, from which we identify

$$\dot{\gamma} = -\frac{dv_z}{dr} \tag{6.29}$$

Assuming $\tau_w > \tau_y$ (i.e., $a > r_c$) and substituting Eqs. (6.26), (6.28), and (6.29) into Casson's equation, Eq. (6.2), yields

$$-\frac{dv_z(r)}{dr} = \frac{1}{2\eta}\left(-\frac{dp}{dz}\right)\left(\sqrt{r} - \sqrt{r_c}\right)^2 \tag{6.30}$$

Using the "no-slip" boundary condition of Eq. (6.20), Eq. (6.30) can be integrated to yield

$$v_z = \frac{1}{4\eta}\left(-\frac{dp}{dz}\right)\left[a^2 - r^2 - \frac{8}{3}r_c^{1/2}\left(a^{3/2} - r^{3/2}\right) + 2r_c(a - r)\right] \quad \text{for } r_c \le r < a \tag{6.31}$$

which reduces to Eq. (6.22) for small r_c. Alternatively, Eqs. (6.27) and (6.28) into Eq. (6.31) give

$$v_z = \frac{a\tau_w}{2\eta}\left\{\left[1 - \left(\frac{r}{a}\right)^2\right] - \frac{8}{3}\sqrt{\frac{\tau_y}{\tau_w}}\left[1 - \left(\frac{r}{a}\right)^{3/2}\right] + \frac{2\tau_y}{\tau_w}\left(1 - \frac{r}{a}\right)\right\} \quad \text{for } r_c \le r < a \tag{6.32}$$

At $r = r_c$ (also $0 \leq r < r_c$), the core velocity is given by Eq. (6.31) as $v_z(r_c)$

$$v_z = \frac{1}{4\eta}\left(-\frac{dp}{dz}\right)\left[a^2 - \frac{8a}{3}\sqrt{r_c a} + 2r_c a - \frac{r_c^2}{3}\right] \quad \text{for } 0 \leq r \leq r_c \tag{6.33}$$

Compare this last result to Eq. (6.24). The flow profile of Eqs. (6.31) and (6.33) is shown in Figure 6.6.

Due to the shape of this profile, r_c is known as the *plug radius*. Using the first equality of Eq. (6.23), we get a complicated equation for the volumetric flow rate, which can be expressed more conveniently as

$$Q = \left[\frac{\pi a^4}{8\eta}\left(-\frac{dp}{dz}\right)\right]F(\varepsilon) \tag{6.34}$$

where

$$F(\varepsilon) = 1 - \frac{16}{7}\varepsilon^{1/2} + \frac{4}{3}\varepsilon - \frac{1}{21}\varepsilon^4, \quad \varepsilon = \frac{2\tau_y/a}{(-dp/dz)} = \frac{\tau_y}{\tau_w} \tag{6.35}$$

Figure 6.7 shows a plot of $F(\varepsilon)$. As $\varepsilon \to 0$, $F(\varepsilon) \to 1$, and Q approaches that of an incompressible Newtonian fluid with $\mu = \eta$. As before, we can calculate

$$v_{\text{MAX}} = v(r_c) \tag{6.36}$$

and

$$v_{\text{avg}} = \left[\frac{a^2}{8\eta}\left(-\frac{dp}{dz}\right)\right]F(\varepsilon) \tag{6.37}$$

the last of which is similar to Eq. (6.25).

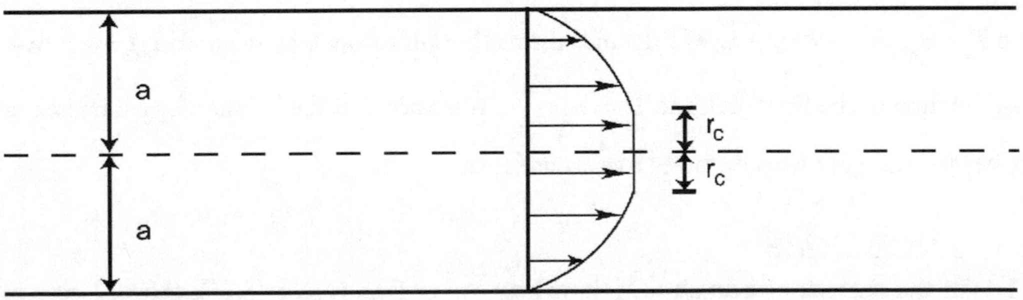

FIGURE 6.6: Velocity profile of a Casson fluid in a circular cylinder. This is a model for blood flow in artery.

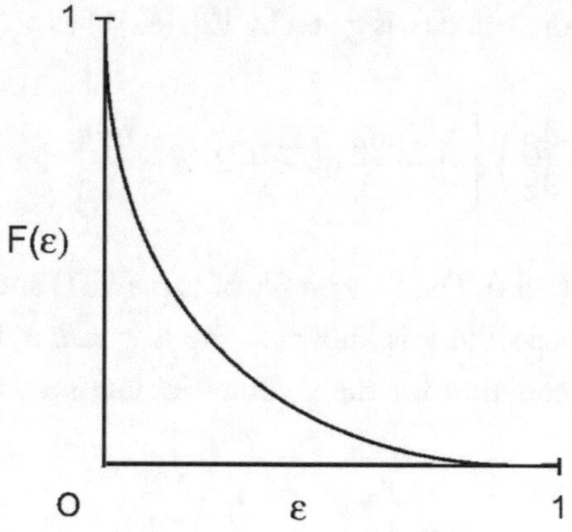

FIGURE 6.7: Plot of $F(\varepsilon)$ versus ε.

This completes the task we set out with, namely, to develop the constitutive equations for blood flow and an equation for the flow rate. Of course, we can increase the complexity of our analysis in several ways. For example, due to the beating of the heart, we have pressure as a function of both time, t, and z. The solution for pulsatile flow, i.e. when

$$\frac{\partial p}{\partial z} = -\frac{\Delta P}{L} + \sum_{n=1}^{N} \left[\psi_n \cos(n\omega t) + \phi_n \sin(n\omega t)\right] = -\frac{\Delta P}{L} + \sum_{n=1}^{N} \mathbb{R}\left[\Upsilon_n e^{n(i\omega t)}\right] \qquad (6.38)$$

of a *Newtonian fluid*, e.g., blood at high shear rates, in a cylindrical tube is [16]

$$v_z(r,t) = \frac{\Delta P}{4\mu L}\left(a^2 - r^2\right) + \mathbb{R}\left[\sum_{n=1}^{N} \frac{i\Upsilon_n}{n\omega\rho}\left(1 - \frac{J_0\left(\frac{i^{3/2}n^{1/2}\alpha r}{a}\right)}{J_0\left(i^{3/2}n^{1/2}\alpha\right)}\right) e^{n(i\omega t)}\right] \qquad (6.39)$$

where $\Upsilon_n = \psi_n - i\phi_n$, $i = \sqrt{-1}$, $\mathbb{R}[\,]$ denotes the real part of the imaginary number, J_0 is a zero-order Bessel function of the first kind, and $\alpha = a\sqrt{\frac{\omega\rho}{\mu}}$. α is known as the Womersley parameter, which describes the ratio of unsteady forces to viscous forces.

6.9 PROBLEMS

1. In the manner of Section 5.9, show how to get Eqs. (6.22)–(6.25). (Hint: Look up the Navier–Stokes equations in cylindrical coordinates.) What do they become when $v_z = v_z(r)$, and $v_r = v_\theta = 0$?

2. Solve for steady Couette flow of a Casson fluid. What is the shear stress at any height above the fixed plate? Find the velocity of the upper plate as a function of the shear stress. (Hint: Reread the demonstration of Section 5.10)
3. Solve for the velocity profile during pressure-driven flow of a Casson fluid between two parallel plates of infinite extent. How does this compare to the solution for a Newtonian fluid? (Hint: One can start with a force balance on a fluid element in Cartesian Coordinates)
4. In this chapter, we solved for the flow in a cylindrical tube of a Newtonian fluid and a Casson fluid. Solve for the flow in a cylindrical tube of a power-law fluid (i.e., find $v_z(r)$) and show that the volumetric flow rate is

$$Q = \frac{n\pi}{3n+1}\left(\frac{\Delta P}{2KL}\right)^{\frac{1}{n}} a^{\frac{(3n+1)}{n}} \tag{6.40}$$

You will need to know that the components of the divergence of a tensor, $\underset{\sim}{A}(r, \theta, z, t)$, in cylindrical coordinates are

$$\begin{aligned}
(\nabla \cdot \underset{\sim}{A})_r &= \frac{\partial A_{rr}}{\partial r} + \frac{1}{r}\frac{\partial A_{r\theta}}{\partial \theta} + \frac{A_{rr} - A_{\theta\theta}}{r} + \frac{\partial A_{rz}}{\partial z} \\
(\nabla \cdot \underset{\sim}{A})_\theta &= \frac{\partial A_{\theta r}}{\partial r} + \frac{1}{r}\frac{\partial A_{\theta\theta}}{\partial \theta} + \frac{A_{r\theta} + A_{\theta r}}{r} + \frac{\partial A_{\theta z}}{\partial z} \\
(\nabla \cdot \underset{\sim}{A})_z &= \frac{\partial A_{zr}}{\partial r} + \frac{1}{r}\frac{\partial A_{z\theta}}{\partial \theta} + \frac{A_{zr}}{r} + \frac{\partial A_{zz}}{\partial z}
\end{aligned} \tag{6.41}$$

The components of the gradient of a scalar, $\alpha(r, \theta, z, t)$, in cylindrical coordinates are

$$(\nabla\alpha)_r = \frac{\partial \alpha}{\partial r} \quad (\nabla\alpha)_\theta = \frac{1}{r}\frac{\partial \alpha}{\partial \theta} \quad \text{and} \quad (\nabla\alpha)_z = \frac{\partial \alpha}{\partial z} \tag{6.42}$$

The gradient of a vector, $\vec{a}(r, \theta, z, t)$, in cylindrical coordinates is

$$[\nabla\vec{a}] = \begin{bmatrix}
\frac{\partial a_r}{\partial r} & \frac{1}{r}\left(\frac{\partial a_r}{\partial \theta} - a_\theta\right) & \frac{\partial a_r}{\partial z} \\
\frac{\partial a_\theta}{\partial r} & \frac{1}{r}\left(\frac{\partial a_\theta}{\partial \theta} + a_r\right) & \frac{\partial a_\theta}{\partial z} \\
\frac{\partial a_z}{\partial r} & \frac{1}{r}\frac{\partial a_z}{\partial \theta} & \frac{\partial a_z}{\partial z}
\end{bmatrix} \tag{6.43}$$

The components of $\underset{\sim}{D}$ in cylindrical coordinates are

$$[\underset{\sim}{D}] = \begin{bmatrix} D_{rr} & D_{r\theta} & D_{rz} \\ D_{\theta r} & D_{\theta\theta} & D_{\theta z} \\ D_{zr} & D_{z\theta} & D_{zz} \end{bmatrix}$$

$$= \begin{bmatrix} \dfrac{\partial v_r}{\partial r} & \dfrac{1}{2}\left(\dfrac{1}{r}\dfrac{\partial v_r}{\partial \theta} + \dfrac{\partial v_\theta}{\partial r} - \dfrac{v_\theta}{r}\right) & \dfrac{1}{2}\left(\dfrac{\partial v_r}{\partial z} + \dfrac{\partial v_z}{\partial r}\right) \\ \dfrac{1}{2}\left(\dfrac{1}{r}\dfrac{\partial v_r}{\partial \theta} + \dfrac{\partial v_\theta}{\partial r} - \dfrac{v_\theta}{r}\right) & \dfrac{1}{r}\dfrac{\partial v_\theta}{\partial \theta} + \dfrac{v_r}{r} & \dfrac{1}{2}\left(\dfrac{\partial v_\theta}{\partial z} + \dfrac{1}{r}\dfrac{\partial v_z}{\partial \theta}\right) \\ \dfrac{1}{2}\left(\dfrac{\partial v_r}{\partial z} + \dfrac{\partial v_z}{\partial r}\right) & \dfrac{1}{2}\left(\dfrac{\partial v_\theta}{\partial z} + \dfrac{1}{r}\dfrac{\partial v_z}{\partial \theta}\right) & \dfrac{\partial v_z}{\partial z} \end{bmatrix} \tag{6.44}$$

5. Solve for the flow of a Bingham plastic in a cylindrical tube. Show that

$$Q = \frac{\pi a^4 \Delta P}{8\mu L}\left[1 - \frac{4}{3}\left(\frac{r_c}{a}\right) + \frac{1}{3}\left(\frac{r_c}{a}\right)^4\right] \tag{6.45}$$

when $r_c < a$. (Hint: Use Eq. 5.53 to define τ_y in terms of the 'plug radius', r_c. Consider flow both within and outside of r_c.)

6. This question is concerned with wall shear stress. Shear stress along the walls of blood vessels has been implicated in atherosclerosis, a disease process in which the body's arteries clog, narrow, and harden.
   (a) What is the mathematical expression for wall shear stress of a Newtonian fluid in a cylinder?
   (b) Evaluate the expression in (a) to find the magnitude of wall shear stress in the left anterior descending (LAD) coronary artery, also known as the window-maker. Assume the LAD has an internal diameter of 3.2 mm, and a pressure drop of 1,300 Pa over 7 mm.
   (c) What is the time-dependent part of the wall shear stress during pulsatile flow? (You will need to know that $\dfrac{d}{dr}[J_0(kr)] = -kJ_1(kr)$, where J_1 is a first-order Bessel function of the first kind.)
7. Assume that the inferior vena cava (IVC) has an inner diameter of 1.5 cm and experiences a pressure drop of 1.0 mmHg over 10 cm. Also assume steady, laminar flow. For blood flowing in the vena cava, $\tau_y = 0.4$ dyne/cm² and $\eta = 4.0$ cP.

(a) Determine the values of ε and $F(\varepsilon)$ for blood flow in the IVC. What does this mean when considering whether to use a Casson fluid or Newtonian fluid to model blood in this situation?

(b) Calculate the flow rate for blood in the IVC assuming it is a Newtonian fluid. Also find the max velocity and sketch the profile.

(c) Repeat part (b), this time modeling blood as a Casson fluid.

8. Read *Hemorheological Observation on 139 Cases of Essential Hypertension by Casson Equation*. Shi *et al.* Clinical Hemorheology 16(4), pp. 559–570, 1996.

   (a) Make a bar graph comparing the yield stress of blood for male and female normal and hypertension subjects. Label the plot clearly, including units, standard deviations, sample size(s), and identification of statistically significant differences where applicable.

   (b) What inference(s) about the usefulness of measuring yield stress can you make?

   (c) Do you agree with discussion point no. 4 (p. 568), specifically the first sentence? Explain, referring to Figures 5–8 on p. 567.

9. Concepts from biomechanics of circulation.

   (a) In the majority of blood vessels in the human body, is blood flow laminar or turbulent?

   (b) As mentioned in class, blood is a thixotropic fluid, meaning its viscosity is time dependent. Although we neglected time dependence in our analysis, how would viscosity actually change with respect to time (i.e. would it increase or decrease)?

   (c) Briefly explain why/how the non-Newtonian behavior of blood flow is attributed to red blood cells.

• • • •

CHAPTER 7
Viscoelasticity

7.1 INTRODUCTION

Thus far we have discussed the material behavior of elastic solids, Newtonian viscous fluids, and some nonlinear viscous fluids. Now we are going to begin discussion of materials that simultaneously exhibit characteristics of both elastic solids and viscous fluids. We mentioned viscoelasticity in the section on nonlinear fluids in Chapter 5, where it was introduced that the current stress in the material depends on aspects of the deformation history. Many materials of interest in biomechanics, including soft tissues, cells, and biological and synthetic polymers, fall into this class of materials and exhibit the following behaviors (see Figures 7.1–7.4):

(a) *Creep*—continued deformation over time of a body maintained under a constant stress
Example: Creep indentation of articular cartilage

(b) *Stress relaxation*—when a body is suddenly strained and the strain is maintained, the stress in the body decreases over time
Example: Stress relaxation of a soft tissue, e.g., tendon, ligament, skin under tension

(c) *Hysteresis*—dissipation of energy during loading–unloading cycles
Example: Cyclic tensile loading of articular cartilage, knee meniscus, or temporomandibular joint disc

(d) *Strain-rate-dependent properties*—when the stress–strain behavior of a material depends on the strain rate
Example: Straight pulls of a soft tissue to a given strain, done at several different strain rates

7.2 DEFINITION OF VISCOELASTICITY

A *viscoelastic material* exhibits creep, stress relaxation, hysteresis, and strain rate-dependent material properties. Thus, viscoelastic constitutive equations provide appropriate models for many tissues and polymers, which are empirically known to exhibit such behaviors.

A viscoelastic material is defined as a material where the stress depends on the strain history of the material. For this reason, viscoelastic materials are often referred to as materials with memory. Mathematically, this definition is described using an integral form, where the integral is known as a *hereditary integral.*

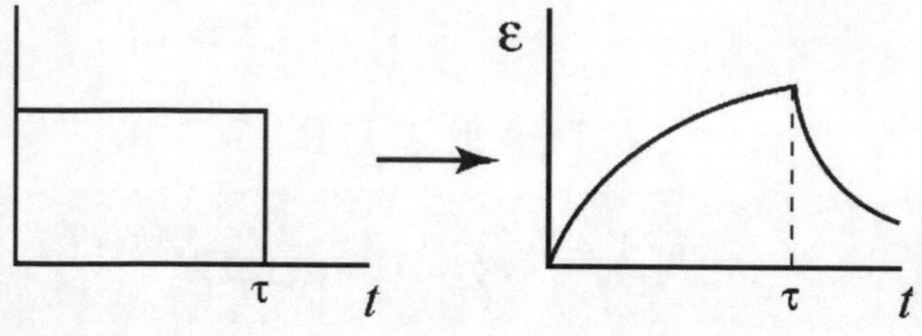

FIGURE 7.1: Creep of a viscoelastic material. A constant stress is applied, and the strain is monitored over time.

An alternative definition is a material where the stress is a function of both the strain and time derivatives of the strain. In this case, differential operators and mechanical circuit equivalents are used to describe viscoelastic materials.

7.3 1-D LINEAR VISCOELASTICITY (DIFFERENTIAL FORM BASED ON MECHANICAL CIRCUIT MODELS)

If one uses the definition that stress in a viscoelastic material depends both on strain and time derivatives of the strain, it is expedient and appropriate to use mechanical circuit models to represent the behavior of simple, idealized viscoelastic materials. Let us first consider the two components that will make up our mechanical circuits. The first is a Hookean spring, which is described by the equation

$$\sigma = E\varepsilon$$

The Hookean spring provides the elastic component of our viscoelastic material. The second component is a dashpot, which is described by

$$\sigma = \eta \frac{\mathrm{d}\varepsilon}{\mathrm{d}t}$$

The dashpot contributes a viscous element to our viscoelastic material. We will describe three canonical spring and dashpot models, and look at how they can be combined to model more complicated behavior.

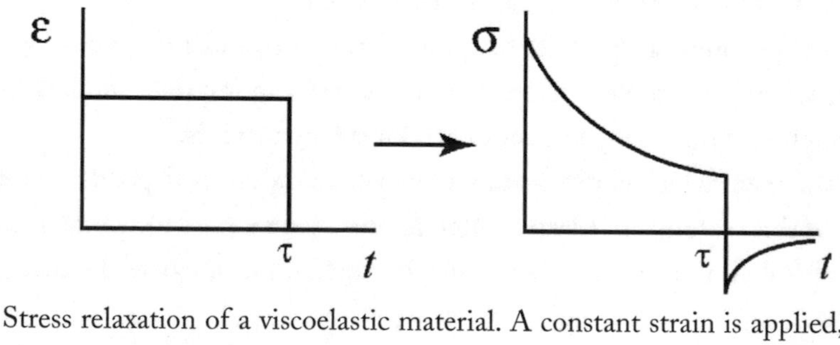

FIGURE 7.2: Stress relaxation of a viscoelastic material. A constant strain is applied, and the stress is monitored over time.

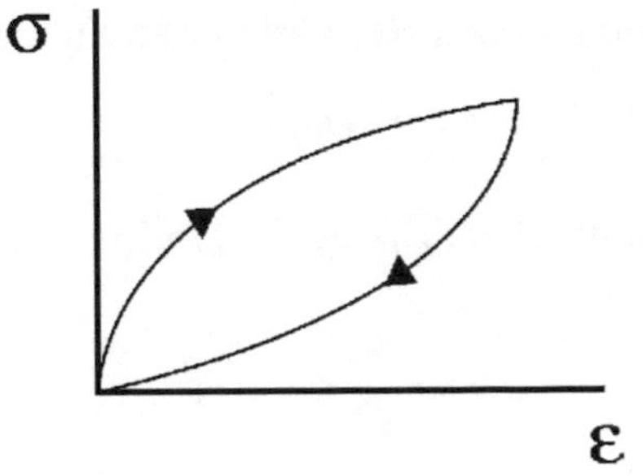

FIGURE 7.3: Hysteresis of stress–strain curve for a viscoelastic material subject to cyclic loading. The loading and unloading curves are not identical, i.e., the same path is not traced.

7.3.1 Maxwell Fluid

Let us consider the simplest mechanical circuit consisting of two components, a spring and dashpot in series. This is known as the *Maxwell fluid*. Because the two components are in series, the stress is the same in both the spring and the dashpot, and the strain is the sum of the individual strains for each element. These relationships can be written as

$$\sigma = E\varepsilon_{\text{spring}} = \eta \frac{\mathrm{d}\varepsilon_{\text{dashpot}}}{\mathrm{d}t} \tag{7.1}$$

and

$$\varepsilon = \varepsilon_{\text{spring}} + \varepsilon_{\text{dashpot}} \tag{7.2}$$

Differentiating Eq. (7.2) and substituting in Eq. (7.1) yields the constitutive equation for a Maxwell fluid

$$\dot{\varepsilon} = \frac{1}{E}\dot{\sigma} + \frac{1}{\eta}\sigma \tag{7.3}$$

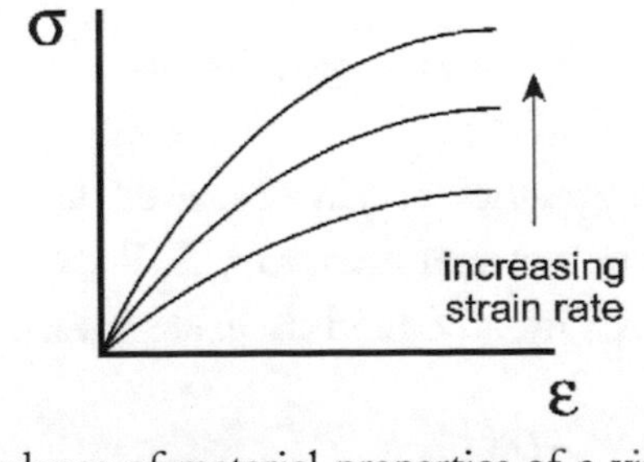

FIGURE 7.4: Strain-rate dependence of material properties of a viscoelastic material. In this graph, stiffness increases with increasing strain rate.

Let us examine the Maxwell fluid during a creep test. From Figure 7.1, we see that, in a creep test,

$$\sigma = \sigma_0 H(t) \tag{7.4}$$

where $H(t)$ is the Heaviside step function. Taking the Laplace transform of Eq. (7.3) we get

$$s\overline{\varepsilon} = \frac{1}{E}s\overline{\sigma} + \frac{1}{\eta}\overline{\sigma} \tag{7.5}$$

where the barred over variable represents the Laplace transform of that variable (e.g.,$\overline{\sigma} = L\{\sigma\}$, where $L\{\ \}$ denotes the Laplace transform). Now, by Eq. (7.4), $\overline{\sigma} = \sigma_0\frac{1}{s}$. Substituting this result into Eq. (7.5), simplifying, and rearranging yields

$$\overline{\varepsilon} = \frac{\sigma_0}{E}\left(\frac{1}{s}\right) + \frac{\sigma_0}{\eta}\left(\frac{1}{s^2}\right) \tag{7.6}$$

The inverse Laplace transform of Eq. (7.6) is

$$\varepsilon(t) = \frac{\sigma_0}{E} + \frac{\sigma_0}{\eta}t \tag{7.7}$$

Thus, when subjected to a step load creep test, the Maxwell fluid will deform unboundedly in a manner linearly dependent on time.

What about stress relaxation? For stress relaxation (see Figure 7.2), we have

$$\varepsilon = \varepsilon_0 H(t) \tag{7.8}$$

so $\overline{\varepsilon} = \varepsilon_0\frac{1}{s}$. Using this expression with Eq. (7.5) and rearranging,

$$\overline{\sigma} = \frac{E\varepsilon_0}{s + \frac{E}{\eta}} \tag{7.9}$$

The inverse Laplace transform of Eq. (7.9) is

$$\sigma(t) = E\varepsilon_0 e^{-(E/\eta)t} \tag{7.10}$$

Thus, when subjected to a step strain, the stress in a Maxwell fluid relaxes back to zero over time in an exponential fashion, with a time constant equal to η/E.

Note that, in each case above, the stress and strain are linearly related. For the stress relaxation response, we can write

$$\frac{\sigma(t)}{\varepsilon_0} = Y(t) = Ee^{-(E/\eta)t} \tag{7.11}$$

where $Y(t)$ is known as the relaxation modulus. Similarly, for the creep response, we find

$$\frac{\varepsilon(t)}{\sigma_0} = J(t) = \frac{1}{E} + \frac{1}{\eta}t \tag{7.12}$$

where $J(t)$ is known as the creep compliance.

While the stress relaxation response of a Maxwell fluid is reasonable and approximates most relaxation behavior seen in viscoelastic materials, the creep response is linear in time, which is in contrast to curves normally acquired from creep test experiments. These behaviors are why this model is known as the Maxwell fluid, namely, continued deformation under a load and complete stress relaxation ("fluid cannot sustain shear stress").

Finally, let us look at what would happen in a so-called standard test where we impose a creep phase followed by a stress relaxation phase (see Figure 7.5). The creep phase is easy. The stress is constant at a value σ_0, and the strain is given by $\varepsilon(t) = \sigma_0 J(t)$. The relaxation phase is more difficult. We define a new variable, $\tau = t - t'$, where t' is the time at which stress relaxation is initiated. We know that at $t = t'$, $\sigma(t') = \sigma_0$. Let $\varepsilon(t') \equiv \varepsilon_1$, where ε_1 is the last value of strain that was reached

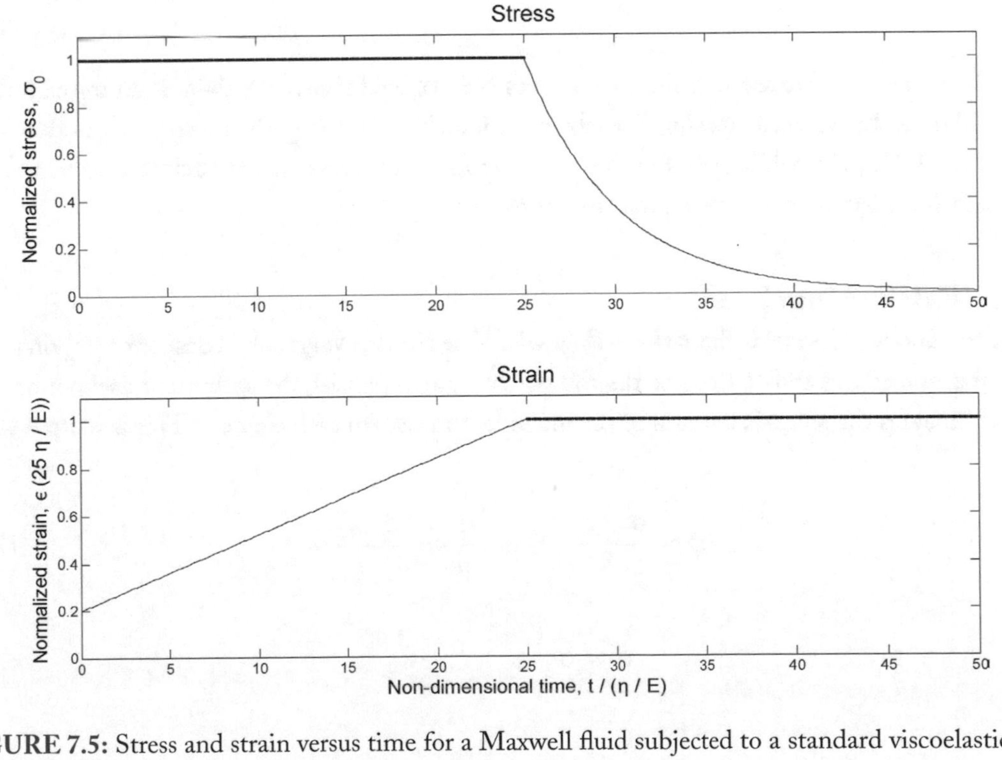

FIGURE 7.5: Stress and strain versus time for a Maxwell fluid subjected to a standard viscoelastic test.

during the creep phase. In terms of the new variable, $\tau = t - t'$, these conditions are $\sigma(0) = \sigma_0$ and $\varepsilon(0) \equiv \varepsilon_1$. Furthermore, for stress relaxation, we want the strain to remain constant at ε_1. We can accomplish this with the Heaviside function shifted in time by t'. Mathematically, $\varepsilon(t) = \varepsilon_1 H(t - t') = \varepsilon_1 H(\tau)$ turns on the strain ε_1 at $t = t'$. The constitutive equation is the same; only the conditions have changed. Thus, from Eq. (7.3), we have $s\bar{\varepsilon} - \varepsilon(0) = \frac{1}{E}s\bar{\sigma} - \frac{1}{E}\sigma(0) + \frac{1}{\eta}\bar{\sigma}$, where we have retained the initial conditions that result when taking the Laplace transform of the first derivative of a function. Let us now solve the problem for times $t \geq t'$, which implies $\tau \geq 0$.

We have $\varepsilon(\tau) = \varepsilon_1 H(\tau) \Rightarrow \bar{\varepsilon} = \frac{\varepsilon_1}{s}$. Making substitutions into the Laplace transformed constitutive equation, $s\frac{\varepsilon_1}{s} - \varepsilon_1 = \frac{1}{E}s\bar{\sigma} - \frac{1}{E}\sigma_0 + \frac{1}{\eta}\bar{\sigma}$. Re-arranging, $\frac{1}{E}\sigma_0 = \left(\frac{1}{E}s + \frac{1}{\eta}\right)\bar{\sigma} = \left(\frac{s\eta + E}{E\eta}\right)\bar{\sigma}$, or $\bar{\sigma} = \left(\frac{E\eta}{s\eta + E}\right)\frac{1}{E}\sigma_0 = \left(\frac{\eta}{s\eta + E}\right)\sigma_0 = \left(\frac{1}{s - (-E/\eta)}\right)\sigma_0$.

Taking the inverse Laplace transform, $\sigma(\tau) = \sigma_0 e^{(-E/\eta)\tau} \Rightarrow \sigma(t) = \sigma_0 e^{(-E/\eta)(t-t')}$. So, for the standard test of a Maxwell fluid we have

$$\sigma(t) = \begin{cases} \sigma_0 & \text{creep, } 0 \leq t < t' \\ \sigma_0 e^{(-E/\eta)(t-t')} & \text{stress relaxation, } t \geq t' \end{cases} \quad \text{and} \quad \varepsilon(t) = \begin{cases} \frac{1}{E} + \frac{1}{\eta}t & \text{creep} \\ \varepsilon_1 & \text{stress relaxation} \end{cases} \tag{7.13}$$

In Figure 7.5, we see that for a step stress (i.e., σ_0 held constant), there is an initial jump in strain followed by strain increasing linearly with time. We then stop the creep phase and hold the current level of strain, while allowing the stress to relax. The stress relaxes back to zero in an exponential fashion based on the relaxation time constant.

7.3.2 Kelvin–Voigt Solid

The next canonical model is the *Kelvin–Voigt solid*. The Kelvin–Voigt solid consists of a spring and a dashpot arranged in parallel. Because the components are in parallel, the spring and dashpot have the same strain, and the stress is the sum of the individual stresses in each element. This is written as

$$\varepsilon = \frac{\sigma_{\text{spring}}}{E} \quad \text{and} \quad \frac{d\varepsilon}{dt} = \frac{\sigma_{\text{dashpot}}}{\eta} \tag{7.14}$$

$$\sigma = \sigma_{\text{spring}} + \sigma_{\text{dashpot}} \tag{7.15}$$

Substituting Eq. (7.14) into (7.15) gives us the constitutive equation for the Kelvin–Voigt solid

$$\sigma = E\varepsilon + \eta\dot{\varepsilon} \tag{7.16}$$

Following the same methodology used for the Maxwell fluid, we can find the stress relaxation and creep responses for the Kelvin–Voigt solid. These are

$$Y(t) = E + \eta\delta(t) \tag{7.17}$$

and

$$J(t) = \frac{1}{E}\left(1 - e^{-(E/\eta)t}\right) \tag{7.18}$$

The creep response of the Kelvin–Voigt solid makes sense. However, the relaxation response consists of an impulse followed by a constant plateau, suggesting instantaneous relaxation (which is typically not observed in solids, though possible in fluids). These behaviors are why this model is described as a solid, namely, limited strain under a finite stress (though the final value is approached gradually, as opposed to instantaneously as would be for an elastic solid => delayed elasticity) and stress relaxation to a residual value.

7.3.3 Standard Linear Solid

One of the most widely used models in linear viscoelasticity is the *standard linear solid*. It consists of a spring and dashpot in series, i.e., the Maxwell fluid, in parallel with another spring (see Figure 7.6). This arrangement is known as the Maxwell form. Arranging a Kelvin–Voigt solid in series with a spring, called the Voigt form of the standard linear solid, yields equivalent behavior (see problem 3).

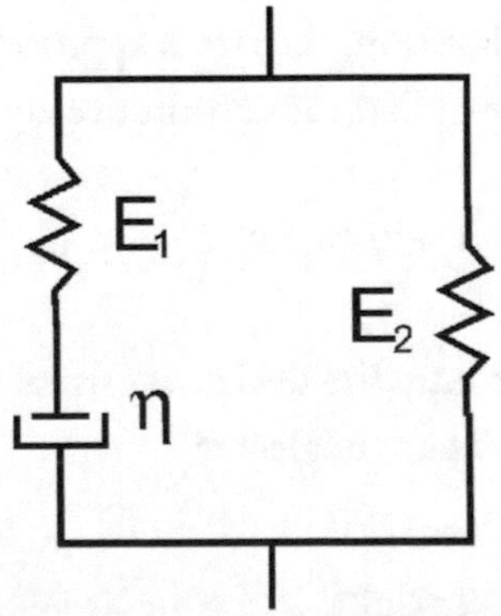

FIGURE 7.6: Standard linear solid (Maxwell form).

Again, the same techniques used to find the constitutive equations for the Maxwell fluid and Kelvin–Voigt solid can be applied here. The results yield the following constitutive equation (see problem 2)

$$(E_1 + E_2)\dot{\varepsilon} + \frac{E_1 E_2}{\eta}\varepsilon = \frac{E_1}{\eta}\sigma + \dot{\sigma} \tag{7.19}$$

The parameters E_1, E_2, and η are typically not reported. Instead, other material constants based on these three parameters are used. The constant E_2 is usually referred to as E_R, or the relaxed modulus. In addition, two time constants, one for stress relaxation (τ_ε) and one for creep (τ_σ), are used. These time constants are given as

$$\begin{aligned} \tau_\varepsilon &= \frac{\eta}{E_1} \\ \tau_\sigma &= \frac{\eta\,(E_1 + E_2)}{E_1 E_2} \end{aligned} \tag{7.20}$$

Note that the creep time constant is larger than the relaxation time constant. Using Laplace transforms, one can solve for the relaxation modulus and creep compliance. They are

$$Y(t) = E_2 + E_1 \mathrm{e}^{-t/\tau_\varepsilon} \tag{7.21}$$

and

$$J(t) = \frac{1}{E_2} - \frac{E_1}{E_2\,(E_1 + E_2)}\mathrm{e}^{-t/\tau_\sigma} \tag{7.22}$$

The standard linear solid model captures behavior typical of many viscoelastic materials, particularly ones of biological interest. We see that, under a step load, the strain response approaches a bounded value, and, during stress relaxation, the stress approaches a residual value.

In invariant form, it can be shown that the constitutive equation for the standard linear solid is

$$\underset{\sim}{T}' + \tau_\varepsilon \dot{\underset{\sim}{T}}' = E_R\left(\underset{\sim}{E}' + \tau_\sigma \dot{\underset{\sim}{E}}'\right) \tag{7.23}$$

In this case, the prime indicates that only the deviatoric stress and strain are being considered, i.e., the contributions of dilatational effects are neglected.

7.3.4 Beyond the Canonical Models

By adding additional elements, we can create models with four or more parameters. For example, two Kelvin–Voigt solid elements in series yield the four-parameter solid, whereas a Voigt form standard linear solid in series with a dashpot yields a four-parameter fluid. Lastly, n Maxwell fluid elements in parallel with each other, and all in parallel with a single spring, yield the *Wiechert model.*

As we have seen, Laplace transforms are very useful in problems based on the differential form of viscoelasticity. Let us further demonstrate this by analyzing the Wiechert model. Because stress sums for elements placed in parallel, the total stress in the Wiechert model is

$$\sigma_{\text{total}} = \sigma_{\text{spring}} + \sum_{i=1}^{n} \sigma_i \tag{7.24}$$

where σ_i is the stress in the ith Maxwell fluid element. The Laplace transform of Eq. (7.24) is

$$\overline{\sigma}_{\text{total}} = \overline{\sigma}_{\text{spring}} + \sum_{i=1}^{n} \overline{\sigma}_i \tag{7.25}$$

Now, the strain in a parallel arrangement is equal in each element. So, $\sigma_{\text{spring}} = E_0\varepsilon$, where E_0 is the stiffness of the lone spring. Hence, $\overline{\sigma}_{\text{spring}} = E_0\overline{\varepsilon}$. The constitutive equation for the ith Maxwell element is, from Eq. (7.3), $\dot{\varepsilon} = \frac{1}{E_i}\dot{\sigma}_i + \frac{1}{\eta_i}\sigma_i$ (no summation on i here). Taking the Laplace transform, and noting that $\sigma(t = 0) = \varepsilon\,(t = 0) = 0$, we get $s\dot{\varepsilon} = \frac{s}{E_i}\overline{\sigma}_i + \frac{1}{\eta_i}\overline{\sigma}_i = \left(\frac{s}{E_i} + \frac{1}{\eta_i}\right)\overline{\sigma}_i$. We can rewrite this as

$$\overline{\sigma}_i = \frac{s}{\left(\frac{s}{E_i} + \frac{1}{\eta_i}\right)}\overline{\varepsilon} = \frac{sE_i}{\left(s + \frac{1}{\tau_i}\right)}\overline{\varepsilon} \tag{7.26}$$

where $\tau_i = \frac{\eta_i}{E_i}$. Making appropriate substitutions, Eq. (7.25) becomes

$$\overline{\sigma}_{\text{total}} = E_0\overline{\varepsilon} + \sum_{i=1}^{n} \frac{sE_i}{\left(s + \frac{1}{\tau_i}\right)}\overline{\varepsilon} = \left(E_0 + \sum_{i=1}^{n} \frac{sE_i}{\left(s + \frac{1}{\tau_i}\right)}\right)\overline{\varepsilon} \tag{7.27}$$

The Laplace plane representation of the time domain constitutive equation is known as the *associated viscoelastic constitutive equation*. Eq. (7.27) is the associated viscoelastic constitutive equation for the Wiechert model. Given a strain profile, we can take the inverse Laplace transform of Eq. (7.27) to solve for stress as a function of time.

Demonstration. A ligament or tendon is pulled in tension at a constant rate of strain. What is the stress response in a Wiechert model containing five Maxwell elements?

Solution. A constant rate of strain is described by $\varepsilon = Rt$. Thus, $\overline{\varepsilon} = \frac{R}{s^2}$. Substituting, and with five elements, Eq. (7.27) becomes $\overline{\sigma}_{\text{total}} = \frac{E_0 R}{s^2} + \sum_{i=1}^{5} \frac{E_i R}{s\left(s + \frac{1}{\tau_i}\right)}$. Taking the inverse Laplace transform yields $\sigma(t) = (E_0 R)t + \sum_{i=1}^{5}(E_i R)\tau_i(1 - \mathrm{e}^{-(t/\tau_i)})$. See reference [17] for an example of the Wiechert model in experimental biomechanics.

7.4 1-D LINEAR VISCOELASTICITY (INTEGRAL FORMULATION)

Our other definition of a viscoelastic material was when the stress depends on the strain history of the material. As we have seen, creep in a linear viscoelastic material can be described by

$$\varepsilon(t) = \sigma_0 H(t) J(t) \tag{7.28}$$

where $J(t)$ is the creep compliance. For a linear material, the strain is directly related to the stress, and the creep compliance is a function of time only. Similarly, the stress relaxation behavior of a linear material can be written as

$$\sigma(t) = \varepsilon_0 H(t) Y(t) \tag{7.29}$$

where $Y(t)$ is the relaxation modulus. Again, stress and strain are linearly related, and the relaxation modulus is a function of time only.

These relationships for creep and stress relaxation assume that a step stress or strain is applied. To find the behavior of a linear viscoelastic material under an arbitrary stress or strain history, we must invoke the *Boltzmann superposition principle*. The Boltzmann superposition principle states that the effect of multiple causes is the sum of the effects of each individual cause. As an example, consider an experiment where a viscoelastic material is instantaneously deformed to some strain and then instantaneously returned to its undeformed state after some arbitrary span of time. The strain function may be written as a superposition of two step functions

$$\varepsilon(t) = \varepsilon_0 \left[H(t) - H(t - t_1)\right] \tag{7.30}$$

Using the Boltzmann superposition principle and Eq. (7.29), the stress can be written as

$$\sigma(t) = \varepsilon_0 \left[Y(t) - Y(t - t_1)\right] \tag{7.31}$$

This can be extended to any arbitrary strain history $\varepsilon(t)$ as shown in Figure 7.7.

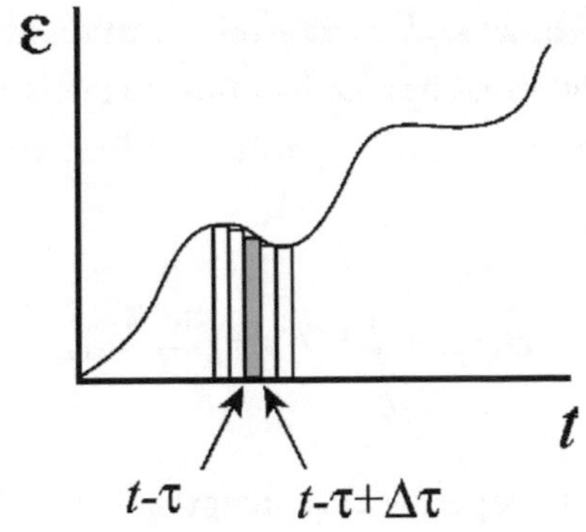

FIGURE 7.7: Arbitrary strain history formed from an infinite sum of step functions.

We will deal with a particular interval of this history, from time $t-\tau$ to $t-\tau+\Delta\tau$, which can be written as the difference of two step functions.

$$\varepsilon(t) = \varepsilon(\tau)\,[H(t-\tau) - H(t-\tau+\Delta\tau)]$$

Using our relationship between stress and strain for stress relaxation under a step strain, as well as the Boltzmann superposition principle, we can write the incremental change in stress due to this strain "pulse" as

$$\mathrm{d}\sigma(t) = \varepsilon(\tau)\,[Y(t-\tau) - Y(t-\tau+\Delta\tau)] \tag{7.32}$$

Given the definition of the derivative,

$$\frac{\mathrm{d}Y(t-\tau)}{\mathrm{d}\tau} = \lim_{\Delta\tau\to 0}\frac{Y(t-\tau+\Delta\tau) - Y(t-\tau)}{\Delta\tau}$$

one can rewrite Eq. (7.32) as

$$\mathrm{d}\sigma(t) = -\varepsilon(\tau)\frac{\mathrm{d}Y(t-\tau)}{\mathrm{d}\tau}\mathrm{d}\tau \tag{7.33}$$

Eq. (7.33) is the incremental change in stress brought about by an infinitesimal strain pulse in the past. If we integrate over time to include all pulses up to the present time t (i.e., the principle of causality), we obtain the equation

$$\sigma(t) = -\int_0^t \varepsilon(\tau)\frac{\mathrm{d}Y(t-\tau)}{\mathrm{d}\tau}\mathrm{d}\tau + Y(0)\varepsilon(t)$$

The second term, $Y(0)\varepsilon(t)$, arises because, for our arbitrary strain history, the final strain is generally not zero. Thus, at the last step, the stress has not had time to relax. The increment of stress resulting from the last step is then given by the second term. The above integral can be cast in the following form

$$\sigma(t) = \int_0^t Y(t-\tau)\frac{d\varepsilon(\tau)}{d\tau}d\tau \tag{7.34}$$

which is known as the Boltzmann superposition integral. The right-hand side of the equation is often referred to as a hereditary, or historical, integral. It represents the contribution of all strains experienced from $t = 0$ to the present. Using the same arguments, one can arrive at a similar integral expressing strain as a function of stress,

$$\varepsilon(t) = \int_0^t J(t-\tau)\frac{d\sigma(\tau)}{d\tau}d\tau \tag{7.35}$$

Based upon Eqs. (7.34) and (7.35), any number of functions could potentially satisfy the general creep and stress relaxation relationships for a linear viscoelastic material. However, these functions are not arbitrary, and there are limitations on the creep and stress relaxation functions. Restrictions are placed upon these functions using fundamental physical principles, most notably the restriction that the rate of energy dissipation is nonnegative, i.e., energy is only conserved (stored) or dissipated, not created.

7.5 3-D LINEAR VISCOELASTICITY

The arguments used in the 1-D integral formulation case can be easily extrapolated to the general 3-D problem, providing the general constitutive equation

$$T_{ij}(t) = \int_0^t C_{ijkl}(t-\tau)\frac{dE_{kl}}{d\tau}d\tau \tag{7.36}$$

Making the assumption of isotropy, Eq. (7.36) reduces to a form analogous to the isotropic linear elastic solid.

$$T_{ij}(t) = \int_0^t \lambda(t-\tau)\delta_{ij}\frac{dE_{kk}}{d\tau}d\tau + \int_0^t 2\mu(t-\tau)\frac{dE_{ij}}{d\tau}d\tau \tag{7.37}$$

*7.6 BOUNDARY VALUE PROBLEMS AND THE CORRESPONDENCE PRINCIPLE

A *boundary value problem* is a problem which has values assigned on the physical boundary of the domain in which the problem is specified. To solve boundary value problems in linear viscoelasticity, we invoke the *correspondence principle*. The correspondence principle states that if a solution of a problem in linear elasticity is known, the solution to the corresponding problem in linear viscoelasticity can be obtained by substituting for the parameters in the linear elasticity solution that can depend on time, a process that generally yields a solution with more ease than from first principles. The correspondence principle is restricted to problems where the boundaries under prescribed loads and displacements *do not change with time*; however, the loads and displacements themselves may be time-dependent.

For 1-D problems, or problems with a constant Poisson's ratio, finding the solution to the viscoelastic counterpart of the elastic problem with the correspondence principle is accomplished as follows: 1. identify which variables can depend on time, 2. take the Laplace transform of the solution and substitute $s\bar{Y}$ or $1/s\bar{J}$ for E_Y, 3. take the inverse Laplace transform to get the viscoelastic solution, and 4. substitute for $Y(t)$ or $J(t)$ from the particular viscoelastic model being used.

Let us look at this process in practice. Recall the example for simple extension of a bar that we previously discussed for linear elasticity. In that example, the solution to the problem, Eq. (4.53), was found to be

$$\varepsilon = \frac{\Delta l}{L_0} = \frac{F}{E_Y A} \tag{7.38}$$

where F is the applied force, Δl is the change in length of the bar, L_0 is the original length of the bar, A is the cross-sectional area of the bar, and E_Y is the Young's modulus of the material. In Eq. (7.38), ε and F can depend on time, whereas L_0 is prescribed. Furthermore, we assume small deformation so that A remains approximately constant. Using the correspondence principle (as described above) and Laplace transforms, we can rewrite Eq. (7.38) as

$$\frac{\Delta l}{s} = \frac{\bar{F} L_0}{s \bar{Y} A} \tag{7.39}$$

The term $\Delta l/s$ arises from the fact that this is a step deformation, and the Laplace transform for the step function $H(t)$ is $1/s$. Rearranging the above equation and taking the inverse Laplace transform, we obtain the equation

$$F(t) = \frac{\Delta l A}{L_0} Y(t) \tag{7.40}$$

where $Y(t)$ is the relaxation modulus. If we consider the viscoelastic material to be a standard linear solid, substituting Eq. (7.21) into Eq. (7.40) yields the final solution for stress relaxation.

$$F(t) = \frac{\Delta l A}{L_0}\left(E_2 + E_1 \mathrm{e}^{-t/\tau_\varepsilon}\right) \tag{7.41}$$

Note that $Y(t)$ for a Maxwell fluid, Kelvin–Voigt solid, or other model could be used if more appropriate for the material of interest. *This has been only a brief introduction to the correspondence principle.* For further information on the correspondence principle, particularly its application in multi-dimensional problems, we direct the reader to the references.

Demonstration. Consider a tendon fibroblast (tenocyte) held at each end by a micropipette. Let us assume that the shape of the cell approximates a circular cylinder with radius a and length L_0. The cell is considered to be a linear viscoelastic solid (modeled with the constitutive equation for a standard linear solid) that is homogeneous, isotropic, and incompressible. Let us assume that we are testing a tendon fibroblast with radius $a = 10\ \mu$m and $L_0 = 50\ \mu$m. The cell is stretched to 10% of its undeformed length ($\Delta l = 5\ \mu$m). The instantaneous load and load at equilibrium are measured and found to be

$$F(0) = 100 \text{ nN}$$
$$F(t \to \infty) = 45 \text{ nN}$$

What are the elastic constants E_1 and E_2?

Solution.

$$F(0) = \frac{\pi a^2 \Delta l}{L_0}(E_2 + E_1) = 100 \text{ nN}$$

$$F(\infty) = \frac{\pi a^2 \Delta l}{L_0} E_2 = 45 \text{ nN}$$

Plugging in the appropriate values for a and L_0 and solving for the elastic constants yields $E_1 = 1.75$ kPa and $E_2 = 1.43$ kPa. Typically, we are interested in the instantaneous and equilibrium behavior of the cell, i.e., if the cell were an elastic material, what would its Young's modulus be immediately after loading and after the cell has come to a quasi-resting state? In this case, the properties would be reported as

Instantaneous modulus = $E_0 = E_1 + E_2 = 3.18$ kPa
Equilibrium (relaxed) modulus = $E_2 = E_R = 1.43$ kPa

In actual experiments, a given solution for the creep or stress relaxation behavior would be fit to all of the available data, rather than focusing solely on the instantaneous and equilibrium responses of the cell. Curve fitting algorithms would be used to obtain the parameters (E_1, E_2, and η) that best fit the experimental data (see problem 15).

7.7 DYNAMIC BEHAVIOR OF VISCOELASTIC MATERIALS

Dynamic or oscillatory testing of viscoelastic materials is a common testing modality. The specimen is subjected to an alternating stimulus while simultaneously measuring the output. We will endeavor to develop the theoretical framework for analyzing viscoelastic materials under dynamic loads. We can write a sinusoidal strain history as

$$\varepsilon(t) = \varepsilon_0 \mathrm{e}^{i\omega t} \tag{7.42}$$

where ω is the angular frequency (in radians per second). We rewrite the Boltzmann superposition integral with a lower limit of $-\infty$ (as sinusoidal functions have no starting point).

$$\sigma(t) = \int_{-\infty}^{t} E(t-\tau)\frac{\mathrm{d}\varepsilon}{\mathrm{d}\tau}\mathrm{d}\tau \tag{7.43}$$

To allow for Eq. (7.43) to explicitly converge, we must decompose the relaxation function $E(t)$ into

$$E(t) = \tilde{E}(t) + E_e \tag{7.44}$$

where E_e is the equilibrium modulus, i.e., $E_e = \lim_{t\to\infty} E(t)$. Combining Eqs. (7.42), (7.43), and (7.44)

$$\sigma(t) = E_e\varepsilon_0 \mathrm{e}^{i\omega t} + i\omega\varepsilon_0 \int_{-\infty}^{t} \tilde{E}(t-\tau)\mathrm{e}^{i\omega\tau}\mathrm{d}\tau$$

If we substitute $t' = t - \tau$ and rewrite the equation using sine and cosine functions, we can obtain the relation

$$\sigma(t) = \varepsilon_0 \mathrm{e}^{i\omega t}\left[E_e + \omega\int_0^{\infty}\tilde{E}(t')\sin\omega t'\mathrm{d}t' + i\omega\int_0^{\infty}\tilde{E}(t')\cos\omega t'\mathrm{d}t'\right] \tag{7.45}$$

Let us now define two frequency-dependent mechanical properties, the *storage modulus*

$$E'(\omega) \equiv E_e + \omega \int_0^\infty \tilde{E}(t') \sin \omega t' \mathrm{d}t' \tag{7.46}$$

and the *loss modulus*

$$E''(\omega) \equiv \omega \int_0^\infty \tilde{E}(t') \cos \omega t' \mathrm{d}t' \tag{7.47}$$

The resultant stress–strain relation can then be written as

$$\sigma(t) = \left[E'(\omega) + iE''(\omega)\right] \varepsilon(t) \tag{7.48}$$

One can see from the form of Eq. (7.48) that the stress is sinusoidal. However, the stress and strain are not in phase. The phase angle, δ, between the stress and strain functions can be found by

$$\tan \delta = \frac{E''(\omega)}{E'(\omega)} \tag{7.49}$$

Eq. (7.49) is referred to as the *loss tangent*. Another mechanical property often cited is the *dynamic modulus*, given by

There is another way to look at dynamic viscoelasticity. A linear relationship exists between stress and strain of the form

$$\sigma(\omega,t) = E^*(i\omega)\varepsilon(\omega,t) \tag{7.50}$$

where periodic loading is considered, and, thus, σ and ε are harmonic functions of (ω,t). The modulus is given by

$$E^*(i\omega) = E'(\omega) + iE''(\omega) \tag{7.51}$$

where $E^*(i\omega)$ is the *complex or dynamic modulus* of the material. Because of the effect of delayed elasticity and viscous flow, σ and ε will, in general, be out of phase. Let us consider a viscoelastic material under a sinusoidal strain,

$$\varepsilon(t) = \varepsilon_0 \sin(\omega t) \tag{7.52}$$

with

$$\sigma(t) = \sigma_0 \sin(\omega t + \delta) \tag{7.53}$$

where ω is the angular frequency and δ is the phase lag. Eq. (7.53) can be rewritten as

$$\sigma(t) = \sigma_0 \sin(\omega t) \cos(\delta) + \sigma_0 \cos(\omega t) \sin(\delta) \tag{7.54}$$

Introducing moduli E' and E'' defined as

$$E' = \frac{\sigma_0}{\varepsilon_0} \cos(\delta) \quad \text{and} \quad E'' = \frac{\sigma_0}{\varepsilon_0} \sin(\delta) \tag{7.55}$$

Eq. (7.55) implies

$$\tan(\delta) = \frac{E''}{E'} \tag{7.56}$$

Eqs. (7.53)–(7.54) into (7.52) results in

$$\sigma(t) = \varepsilon_0 E' \sin(\omega t) + \varepsilon_0 E'' \cos(\omega t) \tag{7.57}$$

Eq. (7.57) says that the σ–ε relationship is defined by E' in phase with ε and E'' which is 90° out of phase with ε. If we use a complex representation of the input and output,

$$\sigma(t) = \sigma_0 e^{i(\omega t + \delta)} \quad \text{and} \quad \varepsilon(t) = \varepsilon_0 e^{i\omega t} \tag{7.58}$$

then

$$E^* = \frac{\sigma}{\varepsilon} = \frac{\sigma_0}{\varepsilon_0} e^{i\delta} = \frac{\sigma_0}{\varepsilon_0}(\cos\delta + i \sin\delta) = E' + iE'' \tag{7.59}$$

where E^* is the complex modulus, E' is the storage modulus (in phase with ε) associated with the energy stored in the specimen due to applied ε, E'' is the loss modulus (out of phase with ε) describing the dissipation of energy due to internal friction, and $\tan\delta$ is the loss tangent.

7.7.1 Dynamic Maxwell Fluid

From Eq. (7.3), we can write the constitutive equation for a Maxwell fluid as

$$\sigma + \lambda\dot{\sigma} = E\lambda\,\dot{\varepsilon} \tag{7.60}$$

where $\lambda = \dfrac{\eta}{E}$. We now let

$$\sigma(t) = \sigma_0 e^{i\omega t} \tag{7.61}$$

Rewriting Eq. (7.59) as $\sigma = \varepsilon\,(E' + iE'')$ yields

$$\varepsilon = \frac{\sigma}{(E' + iE'')} = \frac{\sigma_0 e^{i\omega t}}{(E' + iE'')} \tag{7.62}$$

Now, Eqs. (7.61) and (7.62) into Eq. (7.60) yield

$$E' + iE'' = E\left(\frac{i\omega\lambda}{1 + i\omega\lambda}\right) \tag{7.63}$$

TABLE 7.1: Dynamic viscoelastic properties of the Maxwell fluid and standard linear solid (Maxwell form)

VISCOELASTIC MODEL	STORAGE MODULUS, E′	LOSS MODULUS, E″	LOSS TANGENT, TAN δ
Maxwell fluid	$E\left(\dfrac{\omega^2\lambda^2}{1+\omega^2\lambda^2}\right)$	$E\left(\dfrac{\omega\lambda}{1+\omega^2\lambda^2}\right)$	$\dfrac{1}{\omega\lambda}$
Standard linear solid	$E_2 + \dfrac{E_1\omega^2}{\dfrac{E_1^2}{\eta^2} + \omega^2}$	$\dfrac{\left(\dfrac{E_1^2}{\eta}\right)\omega}{\dfrac{E_1^2}{\eta^2} + \omega^2}$	$\dfrac{\omega\eta E_1^2}{E_1^2E_2 + \omega^2\eta^2(E_1 + E_2)}$

Writing the right-hand side of Eq. (7.63) as a complex number (i.e., $a + ib$) yields the storage modulus, loss modulus, and loss tangent for a Maxwell fluid. Table 7.1 gives the dynamic viscoelastic properties of the Maxwell fluid and Maxwell form of the standard linear solid. Problem 12 is reserved for the Kelvin–Voigt solid.

7.8 LIMITING CASES OF LINEAR VISCOELASTICITY ARE THE HOOKEAN SOLID AND NEWTONIAN VISCOUS FLUID

We would like to take a brief moment to elucidate how linear viscoelasticity relates to linear elasticity and Newtonian viscous fluids. From our discussion of linear elasticity, we know there is no time dependence in the constitutive equations. Thus, to arrive at linear elasticity from linear viscoelasticity, we make the relaxation functions independent of time. Thus, Eq. (7.37) becomes

$$\begin{aligned} T_{ij}(t) &= \lambda\delta_{ij}\int_0^t \frac{dE_{kk}}{d\tau}d\tau + 2\mu\int_0^t \frac{dE_{ij}}{d\tau}d\tau = \lambda\delta_{ij}\int_0^t dE_{kk} + 2\mu\int_0^t dE_{ij} \\ &= \lambda\delta_{ij}\, E_{kk}\big|_0^t + 2\mu\, E_{ij}\big|_0^t = \lambda\delta_{ij}E_{kk}(t) + 2\mu E_{ij}(t) \end{aligned} \tag{7.64}$$

where we have assumed an unstrained state at $t = 0$. Thus, at each time t, the constitutive equation is the same as for linear elasticity, Eq. (4.30).

To arrive at a Newtonian viscous fluid, we set the relaxation functions equal to $\lambda\delta(t-\tau)$ and $2\mu\delta(t-\tau)$, where $\delta(t-\tau)$ is the Dirac delta function. Then, Eq. (7.37) becomes

$$\begin{aligned} T_{ij}(t) &= \lambda\delta_{ij}\int_0^t \delta(t-\tau)\frac{dE_{kk}}{d\tau}d\tau + 2\mu\int_0^t \delta(t-\tau)\frac{dE_{ij}}{d\tau}d\tau \\ &= \lambda\delta_{ij}\frac{dE_{kk}(t)}{dt} + 2\mu\frac{dE_{ij}(t)}{dt} = \lambda\delta_{ij}D_{kk}(t) + 2\mu D_{ij}(t) \end{aligned} \tag{7.65}$$

where we have used the fact that the time derivative of the strain tensor is the strain rate tensor under infinitesimal displacements [18]. Thus, at each time t, we recover the viscous part of the constitutive equation for a Newtonian fluid, Eq. (5.17).

Finally, it should be noted that the results of this chapter apply to infinitesimal deformations. Extensions of this theory, such as Y.C. Fung's QLV (quasi-linear viscoelasticity) theory [14] and hyperviscoelasticity [7], should be employed when finite strains prevail. For further discussion of viscoelasticity, we refer the interested reader to references [19-21].

7.9 PROBLEMS

1. The springs and dashpots introduced in this chapter have analogs in circuit theory.
   (a) Show that the equivalent stiffness of two springs in parallel is $E_{\text{equiv}} = E_1 + E_2$.
   (b) Show that two springs in series have an equivalent stiffness given by $\frac{1}{E_{\text{equiv}}} = \frac{1}{E_1} + \frac{1}{E_2}$.

   These two expressions are similar to capacitors in an electrical circuit.
2. In this problem you will more formally investigate the Maxwell form of the standard linear solid model.
   (a) Derive the constitutive equation for the standard linear solid, Eq. (7.19).
   (b) Choose one of the following: (i) Calculate $Y(t)$ or (ii) calculate $J(t)$ for the standard linear solid subjected to a step strain or step stress, respectively.
   (c) Suppose the strain or stress history is arbitrary, i.e., not a step displacement or load. How would you calculate the stress or strain as a function of time for the arbitrary history? (Hint: Reread Section 7.4.)
3. The constitutive equation for the Maxwell form of the standard linear solid, Eq. (7.19), can be written as $\sigma + C_3\dot{\sigma} = C_1\dot{\varepsilon} + C_2\varepsilon$, where the C_i's are "lumped parameters." Show that the Voigt form of the standard linear solid (a Kelvin–Voigt solid in series with a spring) can be written the same way, though the lumped parameters will be a collection of different constants. Note that, though the lumped parameters are a collection of different constants, the governing equations are the same, yielding identical material behavior!
4. Show that the three-parameter fluid (a Kelvin–Voigt solid in series with a dashpot or a Maxwell fluid element in parallel with a dashpot) has the constitutive equation of the form $\sigma + C_3\dot{\sigma} = C_1\dot{\varepsilon} + C_2\ddot{\varepsilon}$. What are the C_i's in terms of the E and η's? Obtain the strain as a function of time during a creep test. Based on the results of the creep test, how does this model represent fluid behavior?
5. Solve for stress relaxation of the Wiechert model when subjected to a step strain, $\varepsilon = \varepsilon_0 H(t)$. What is the relaxation modulus? This particular relaxation modulus is known as the "Prony series" expansion. What is the response of the Wiechert model to an arbitrary strain history?

6. Show that $\int_0^t J(t-\tau)\,Y(\tau)d\tau = \int_0^t Y(t-\tau)\,J(\tau)d\tau = t$, where $J(t)$ and $Y(t)$ are the creep compliance and relaxation moduli, respectively. (Hint: What is the relationship of $J(t)$ and $Y(t)$ in the Laplace domain?)
7. Write the 3-D integral form of the constitutive equation for
   (a) An orthotropic material
   (b) A transversely isotropic material
   (Hint: Recall Section 4.11)
8. A beam is simply supported at both ends, and a step load ($F = F_0 H(t)$) is applied perpendicular to the beam at its midspan. The beam is made of a material that can be adequately modeled as a Kelvin–Voigt solid. Plot the displacement versus time at the point of max deflection (x_2^{max}). The elastic solution for the deflection of a beam of length L, elastic modulus E_Y, and cross-sectional moment of inertia I subject to this loading is

$$x_2(x_1) = \frac{F}{48E_Y I}\left(4x_1^3 - 3L^2 x_1\right)$$

   What about the graph indicates "solid-like" behavior?
   (Hint: Use an intermediate relationship you get in problem 6.)
9. Using the correspondence principle, find the deflection of a standard linear solid beam, $\theta(t)$, under pure bending. Evaluate your expression for an applied moment of $M(t) = ct$.
10. Consider indentation of a soft, hydrated tissue. The material is modeled as a semi-infinite elastic solid indented with a circular cylindrical punch. The elastic solution for this problem relating applied force to displacement is

$$F = \frac{E_Y}{1-\nu^2}(2a)\,d$$

   where F is the applied force, a is the radius of the indenter, and d is the indentation depth.
   (a) Assuming that $\nu = 0.5$ (i.e., the material is incompressible), find the indentation depth as a function of time, $d(t)$, for a step load $F(t)=F_0H(t)$ applied to a standard linear solid. (Hint: Recall $J(s)Y(s)=1/s^2$)
   (b) Using your solution from part (a), plot $d(t)$ versus t. What is the displacement immediately upon loading? At equilibrium?
   See reference [22] for an example of viscoelastic testing of cells and the correspondence principle.
11. A proteoglycan solution can be modeled as a Maxwell fluid.

(a) Verify the storage modulus (E'), loss modulus (E''), and loss tangent (δ) given in Table 7.1.

(b) Sketch a plot of the loss tangent versus frequency. How does the material behave at high frequencies?

12. Find the storage modulus (E'), loss modulus (E''), and loss tangent (δ) for the Kelvin–Voigt solid.

13. In this problem, we will look at energy dissipation during cyclic loading of a viscoelastic material. Consider a sinusoidal stress given by $\sigma(t) = \sigma_0 \sin(\omega t)$.

(a) Using Eq. (7.57), what is the in-phase strain? Out-of-phase strain? From basic physics, the work done is equal to the integral of force dot product with displacement, i.e., $W = \int \vec{F} \cdot d\vec{l}$. Show that the work, equivalent to energy through the work-energy principle, per volume can be written $\frac{W}{V} = \int \sigma \frac{d\varepsilon}{dt}\, dt$.

(b) Evaluate the energy per volume for a single loading cycle, i.e., integrate from 0 to $2\pi/\omega$, to show that energy is conserved by in-phase strain, but not by out-of-phase strain. This result is why viscoelastic materials exhibit hysteresis. The out-of-phase energy is lost to the environment as heat.

14. Use the graph shown in Figure 7.8 to plot the behavior of the Kelvin–Voigt and standard linear solids when subjected to the standard test described in the section on the Maxwell fluid. Comment on the instantaneous and equilibrium behaviors.

15. In the tenocyte demonstration, we found the instantaneous and relaxation modulus during a stress relaxation experiment. Use the following data to find the viscosity, η, of this tenocyte. Matlab©'s "cftool" command is one possible method for doing this.

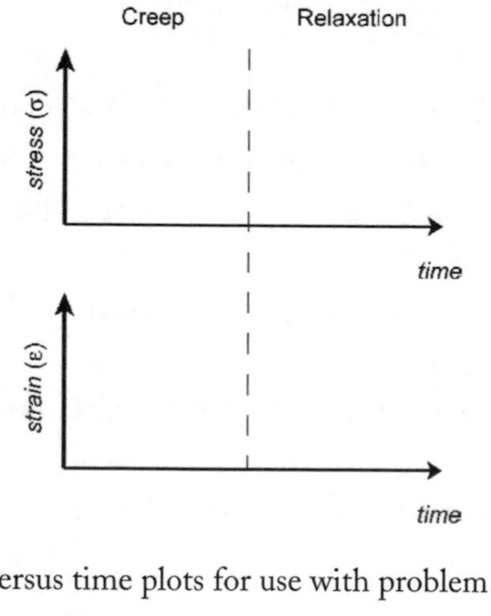

FIGURE 7.8: Stress and strain versus time plots for use with problem 14.

TIME (s)	FORCE (nN)
0	100
0.65	95.5
1.3	91.5
2	87.8
2.5	84.4
4.5	75.7
6.5	68.9
9.75	60.8
13	55.4
16.25	51.8
25	46.9
40	45.3
55	45.15
80	45
90	44.9
155	45
220	44.95
250 ($\approx \infty$)	45

• • • •

CHAPTER 8

*Poroelasticity and Thermoelasticity

8.1 INTRODUCTION

This chapter is concerned with coupled theories, in particular poroelasticity and thermoelasticity. A coupled theory is a theory in which field variables interact with each other. For example, in poroelasticity, the strain field inside a material affects the fluid pressure and solid stress. Thus, a change in fluid pressure can effect a change in solid stress. In thermoelasticity, the temperature field and stress field affect one another, such that a change in temperature can cause a change in stress or vice versa. These theories allow an expanded description of the material's behavior under different conditions. For example, we can consider what happens to the fluid pressure in a tissue when it is loaded or how changing the temperature can induce strain. However, these advancements comes at the expense of somewhat more complicated mathematics. Poroelasticity has been used to describe several biological tissues, such as bone, cartilage, and the intervertebral and temporomandibular joint discs, while thermoelasticity has been more widely used with biomaterials.

8.2 POROELASTICITY

8.2.1 Terzaghi's Principle of Effective Stress

Poroelasticity is the study of the behavior of fluid-saturated elastic porous media. A porous material is a solid containing an interconnected network of pores (voids) filled with a fluid (liquid or gas). The solid matrix and the pore fluid are assumed to be continuous, forming two interpenetrating continua (e.g., a sponge). Materials, both naturally occurring (i.e., rocks, soils, biological tissues) and synthetic (e.g., foams and ceramics), have been described with poroelastic theory.

In poroelasticity, the principle of effective stress is ascribed to Terzaghi, who developed it while performing pioneering work on the 1-D consolidation of soils. A fully fledged theory of poroelasticity was not given until 1941 by Biot [23]. As applied to poroelasticity, the principle of effective stress states that the total stress at a given point in the material is the sum of the solid stress and fluid pressure. Mathematically (recall we define the normal stress as $-p$ in a fluid),

$$\underset{\sim}{T} = \underset{\sim}{T}' - p\underset{\sim}{I} \tag{8.1}$$

where $\underset{\sim}{T}$ is the total stress, $\underset{\sim}{T}'$ is the effective stress (i.e., stress in the solid matrix), and p is the hydrostatic pressure. Though we will restrict our attention in this chapter to linear poroelasticity, note that choosing $\underset{\sim}{T}'$ to be given by a hyperelastic or viscoelastic constitutive equation is also possible, though the latter is called poroviscoelasticity.

8.2.2 Darcy's Law

Darcy's law is a constitutive equation describing the flow of fluid through a porous medium. It was empirically deduced by Darcy in the latter part of the 19th century during his experiments on the flow of water through sand. It has since been derived theoretically [11]. The theoretical analyses have shown that, though having limitations, Darcy's law is valid for Newtonian liquids at low velocities. Thus, Darcy's law has often been employed in the study of the flow of biological fluids through tissues. Darcy's law says that the fluid flux is directly proportional to the permeability of the material through which the fluid is flowing and the pressure gradient driving the flow, while it is inversely proportional to the fluid's viscosity. Mathematically,

$$\vec{q}_{\text{fluid}} = -\frac{1}{\mu_{\text{fluid}}} \underset{\approx}{\kappa} \left(\nabla p - \rho g \vec{e}_3 \right) \tag{8.2}$$

where the minus sign ensures that fluid flows "downhill," i.e., from high to low pressure. The tensor $\underset{\sim}{\kappa}$ is the anisotropic permeability tensor, reflecting the fact that the permeability of the solid through which the fluid is flowing need not be the same in all directions. The matrix representation of $\underset{\sim}{\kappa}$ is

$$[\underset{\approx}{\kappa}] = \begin{bmatrix} \kappa_{11} & \kappa_{12} & \kappa_{13} \\ \kappa_{12} & \kappa_{22} & \kappa_{23} \\ \kappa_{13} & \kappa_{23} & \kappa_{33} \end{bmatrix} \tag{8.3}$$

If the material is isotropic with respect to permeability, $\underset{\sim}{\kappa} = \kappa \underset{\sim}{I}$, where κ is a constant. The term $\rho g \vec{e}_3$ enters, as flow is not affected by the vertical pressure drop caused by gravity under hydrostatic conditions. By subtracting the gravitational pressure drop from the total pressure drop, we obtain the pressure contributing solely to flow. Furthermore, we see that when $\nabla p \sim \Delta P / L \gg \rho g$, gravitational effects can be neglected. Finally, fluid velocity within the pores (i.e., seepage velocity) is related to the flux by the porosity. The fluid velocity is

$$\vec{v} = \frac{\vec{q}_{\text{fluid}}}{\phi} \tag{8.4}$$

where $\phi = \frac{V_{\text{fluid}}}{V_{\text{total}}}$ is the porosity of the material. The flux is divided by porosity to account for the fact that only a fraction of the total volume is available for flow.

Demonstration. An incompressible fluid of viscosity μ is flowing through a porous medium of permeability $\kappa \underset{\sim}{I}$ and length L. The pressure on one side of the medium is held at P_1, while on the other side it is held at P_2. What is the velocity of the fluid within the medium?
Solution. Let us align the direction of flow with the x_1 axis and assume gravitational effects can be neglected. As the fluid is incompressible, the continuity equation is given by Eq. (5.6), or $\nabla \cdot \vec{v} = 0$. From Eqs. (8.4) and (8.2) we get

$$\nabla \cdot \vec{v} = \frac{1}{\phi} \nabla \cdot \vec{q}_{\text{fluid}} = \frac{1}{\phi} \nabla \cdot \left[-\frac{1}{\mu_{\text{fluid}}} \kappa \underset{\sim}{I} (\nabla p) \right] = -\frac{\kappa}{\phi \mu_{\text{fluid}}} \nabla \cdot (\nabla p) = -\frac{\kappa}{\phi \mu_{\text{fluid}}} \nabla^2 p = 0$$

where the identity tensor was eliminated via Eq. (1.46). Assuming the pressure depends only on x_1, the above expression becomes

$$\frac{\mathrm{d}p}{\mathrm{d}x_1} = 0$$

Solving the above equation subject to the pressure boundary conditions yields $p = \left(\frac{P_2 - P_1}{L} \right) x_1 + P_1$. The velocity comes from back substitution into Eqs. (8.2) and (8.4). The velocity is

$$\vec{v} = \frac{1}{\phi} \left(\frac{P_1 - P_2}{L} \right) \frac{\kappa}{\mu_{\text{fluid}}} \vec{e}_1$$

8.2.3 Constitutive Equations and Material Constants

Having discussed the principle of effective stress and Darcy's law, let us now turn to the constitutive equations of linear poroelasticity. There are several ways to formulate the constitutive equations of poroelasticity depending on the choice of independent variables. In this book, we present the *mixed stiffness formulation.* "Stiffness" refers to stress being one of the dependent variables, and mixed refers to using the increment of fluid content as the other (as opposed to pressure, in which case it would be a *pure stiffness formulation*) [24]. *Increment of fluid content,* ζ, is the third, and final, new concept we will present with respect to poroelasticity. It is equal to the volume of fluid added to or removed from a control volume. Thus, increment of fluid content is essentially the "fluid strain", as it is the dimensionless ratio of two volumes.

For stress in a linear anisotropic poroelastic continuum,

$$T_{ij} = C_{ijkl} E_{kl} - \alpha_i p \delta_{ij} \tag{8.5}$$

Note: There is no summation on i in the last term. Thus, there are three independent α_is, one for each direction. There is no contribution of the pressure term to the shear stresses because, as we have learned, fluids cannot support a shear stress. We will define α momentarily. If we consider an isotropic material, Eq. (8.5) becomes

$$T_{ij} = 2\mu_{\text{solid}} E_{ij} + \lambda E_{kk}\delta_{ij} - \alpha p \delta_{ij} \tag{8.6}$$

Comparing Eqs. (8.6) and (8.1), we see that the effective stress in an isotropic, linear poroelastic material is given by the familiar expression for stress in a linearly elastic solid. In linear elasticity, we had nine equations with nine unknowns (see Chapter 4). Introduction of pressure into the constitutive equation leads to an additional unknown. As it was for fluids, the additional equation we will use is the continuity equation. But, before we see how the system of equations comes together, we must also introduce the constitutive equation for the increment of fluid content.

In the mixed stiffness formulation, the increment of fluid content for isotropic materials is

$$\zeta = \alpha E_{kk} + \frac{\alpha}{K_u B} p \tag{8.7}$$

In these equations

$$\alpha \equiv \left.\frac{\partial \zeta}{\partial E_{kk}}\right|_{p=\text{constant}} = \frac{K}{H} \tag{8.8}$$

The coefficient α is known as the *Biot–Willis coefficient*. It is the ratio of fluid added to the change in *bulk* volume (recall E_{kk} is the dilatation). Furthermore, K is the drained bulk modulus ($\lambda + {}^2\!/_3\mu$), and $1/H$ is the poroelastic expansion coefficient. The latter is a measure of the change in increment of fluid content relative to a change in the average applied stress at constant pore pressure. For reference, unconstrained means the stress (T) is constant, constrained means the strain (E) is constant, undrained means there is no change in fluid content within a control volume ($\zeta = 0$), and drained means the pressure (p) is constant.

Next, we have *Skempton's coefficient*, B. It is given as

$$B \equiv -\left.\frac{\partial p}{\partial T}\right|_{\zeta=\text{constant}} = \left.\frac{\partial E_{kk}}{\partial \zeta}\right|_{T=\text{constant}} = \frac{R}{H} \tag{8.9}$$

Skempton's coefficient is a measure of load distribution between the solid and the fluid. It must be a number between zero and one, tending toward one for completely saturated materials (i.e., load is shared by pores filled completely with fluid) and toward zero for void space pores (e.g., highly compressible gases), as the solid framework must support the load. Alternatively, Skempton's coefficient is the ratio of volumetric strain to increment of fluid content in unconstrained conditions.

Furthermore, $1/R \equiv S_T$ is the *unconstrained specific storage coefficient*. For an incompressible fluid, $1/R \to 0$, as a large increase in fluid pressure will cause an insignificant increase in fluid volume.

Finally, K_u is the *undrained bulk modulus*.

$$K_u \equiv \left.\frac{\partial T}{\partial E}\right|_{\zeta=\text{constant}} = \frac{K}{1-\alpha B} \tag{8.10}$$

The bulk moduli have their familiar interpretation of the change in volume under hydrostatic pressure. Combining α, K_u, and B as done in Eq. (8.7) yields the *constrained specific storage coefficient*, S_E.

$$S_E \equiv \left.\frac{\partial \zeta}{\partial p}\right|_{E_{kk}=\text{constant}} = \frac{1}{M} = \frac{\alpha}{K_u B} \tag{8.11}$$

The specific storage coefficients measure the ratio of increment of fluid content to a change in pore pressure under different conditions. Thus, there are four material coefficients needed to fully describe an isotropic poroelastic material, μ, λ, α, and S_E. There are many more poroelastic material coefficients, and descriptions thereof, depending on the formulation of the constitutive equations used. Many also have micromechanical interpretations.

Assuming the material properties have been measured, the unknowns in this system are 11: six stress components, three displacement components (yielding six strains), one pore pressure, and one increment of fluid content. Hence, we need 11 equations: seven constitutive equations (six stress and one increment of fluid content), three force equilibrium equations (one for each direction), and one fluid continuity equation (uses Darcy's Law and, in general, becomes a nonhomogeneous "Heat" equation in pressure, even in the absence of a traditional source).

The system of equations is solved elastostatically (i.e., assume static equilibrium at each instant of time). Wave propagation is ignored in this quasi-static approximation. Thus, the system passes through a sequence of equilibrium states; however, this path is irreversible due to frictional drag loss as the fluid flows past the solid matrix. As will be shown below, using a displacement–pressure formulation results in a system of four coupled equations (by substituting the strain–displacement, Darcy's law, and the constitutive relationships into the conservation equations) with the unknowns being three displacements and one pore pressure. Coupling of the equations comes from the pressure (p) term in the equations of force equilibrium and from the volumetric strain (E_{kk}) in the fluid flow equation.

8.2.4 *u–p* Formulation of Poroelastic Governing Equations

Let us first look at the stress equilibrium equations. Using Eqs. (3.21) and (8.6),

$$\frac{\partial}{\partial x_j}\left(2\mu_{\text{solid}}E_{ij} + \lambda E_{kk}\delta_{ij} - \alpha p\delta_{ij}\right) + \rho B_i = \rho a_i \tag{8.12}$$

Assuming quasi-static conditions, we obtain

$$\mu_{\text{solid}} \frac{\partial^2 u_i}{\partial x_k^2} + (\lambda + \mu_{\text{solid}}) \frac{\partial^2 u_k}{\partial x_i \partial x_k} = \alpha \frac{\partial p}{\partial x_i} - \rho B_i \tag{8.13}$$

Recall that k is summed on in the expression ∂x_k^2. This expression shows the coupling between displacement and pressure. Eq. (8.13) should remind you of Navier's equations. Taking the x_3 direction as an example, Eq. (8.13) is

$$\mu_{\text{solid}} \left(\frac{\partial^2 u_3}{\partial x_1^2} + \frac{\partial^2 u_3}{\partial x_2^2} + \frac{\partial^2 u_3}{\partial x_3^2} \right) + (\lambda + \mu_{\text{solid}}) \left[\frac{\partial^2 u_1}{\partial x_3 \partial x_1} + \frac{\partial^2 u_2}{\partial x_3 \partial x_2} + \frac{\partial^2 u_3}{\partial x_3^2} \right] = \alpha \frac{\partial p}{\partial x_3} - \rho B_3 \tag{8.14}$$

Now we will examine the fluid continuity equation. Heuristically, the time rate of change in increment of fluid content in a control volume is equal to any source producing fluid in the control volume (Q) less the divergence of the fluid from the control volume. Mathematically,

$$\frac{\partial \zeta}{\partial t} = -\nabla \cdot \vec{q}_{\text{fluid}} + Q\left(\vec{x}, t\right) \tag{8.15}$$

Note that a negative Q would correspond to a fluid sink. In Eq. (8.15), Q is a traditional source term having units of (volume fluid contributed/reference volume · time) and $\vec{q}$ is the flux vector (m/s). The gradient operator, ∇, adds 1/length to the dimensions such that the divergence of the flux is 1/time, agreeing with the time derivative of the increment of fluid content. We now substitute Darcy's law, neglecting gravity and with κ and μ_{fluid} independent of spatial location, into the fluid continuity equation to obtain

$$\frac{\partial \zeta}{\partial t} = -\nabla \cdot \left(-\frac{\kappa}{\mu_{\text{fluid}}} \nabla p \right) + Q = \frac{\kappa}{\mu_{\text{fluid}}} \nabla^2 p + Q = k \nabla^2 p + Q \tag{8.16}$$

Recall that κ is the intrinsic permeability of solid, and μ is the fluid viscosity. The ratio, $\kappa / \mu_{\text{fluid}} = k$, is the hydraulic conductivity and has units $m^4/N{\cdot}s$. The Laplacian, ∇^2, adds $1/\text{length}^2$ to the dimensions such that the Laplacian of the pressure is N/m^4.

Substitution of Eq. (8.7) with constant material coefficients into Eq. (8.16) yields

$$\alpha \frac{\partial E_{kk}}{\partial t} + \frac{\alpha}{K_u B} \frac{\partial p}{\partial t} = k \nabla^2 p + Q \tag{8.17}$$

This expression can be rearranged to

$$\frac{\partial p}{\partial t} = \frac{K_u B}{\alpha} k \nabla^2 p + \left[\frac{K_u B}{\alpha} \left(Q - \alpha \frac{\partial E_{kk}}{\partial t} \right) \right] \tag{8.18}$$

Eq. (8.18) should be recognized as the "Heat" equation in p with a source term. Here, the source term consists of everything within the brackets.

This completes the u–p formulation of poroelasticity. Eqs. (8.13) and (8.18) represent four equations for the four unknowns, u_1, u_2, u_3, and p. They form a system of coupled PDEs that must be solved simultaneously subject to appropriate initial and boundary conditions on the displacements and pressure. These equations can be cast in many different forms, depending on the choice of independent variables. We refer the reader to reference [24] for more details.

The above equations, Eqs. (8.13) and (8.18), require six independent material coefficients, λ, μ_{solid}, k, α, B, and K_u. If we assume the solid matrix is intrinsically incompressible, $\alpha = 1$. Furthermore, for an infinitely *incompressible* fluid, $B = 1$ and $K_u = \infty$. Taken together, these assumptions mean $\alpha / K_u B = S_E = 0$. The assumptions of the incompressible nature of both the solid and fluid components are justified for most tissues and, hence, are often employed in biomechanics. These assumptions leave only three material coefficients, λ, μ_{solid}, k. Under these assumptions, Eq. (8.13) becomes

$$\mu_{\text{solid}} \frac{\partial^2 u_i}{\partial x_k^2} + (\lambda + \mu_{\text{solid}}) \frac{\partial^2 u_k}{\partial x_i \partial x_k} = \frac{\partial p}{\partial x_i} - \rho B_i \tag{8.19}$$

and Eq. (8.17) becomes

$$\frac{\partial E_{kk}}{\partial t} = k \nabla^2 p + Q \tag{8.20}$$

By Eq. (8.16), Eq. (8.20) is the statement that the time rate of change of the dilatation is equal to the time rate of change of the increment of fluid content. This makes sense when considering both the solid and fluid to be incompressible, as any movement of an incompressible fluid into a space must correspondingly increase the volume (i.e., dilatation) of that space. It is important to recognize that the assumption of incompressible solid and fluid components does not mean loading will not cause deformation of the tissue. Indeed, compressive loading leads to relative motion between the solid and fluid, whereby the pore size changes as fluid is effluxed from the material due to pressure gradients. Concomitantly, the solid network experiences a collapse in size as it comes to occupy the space previously occupied by the fluid.

8.2.5 Consolidation of a Finite Layer (Terzaghi's Problem)

As an example of poroelasticity, we will investigate the consolidation of a finite layer consisting of a solid matrix inundated with a pore fluid. We confine the finite layer of height L on five of six sides within a rectangular parallelepiped and apply a load through a permeable platen on the free surface. Our goal is to obtain the pressure and displacement fields within the finite layer when subjected to a load.

The above problem description amounts to the condition of uniaxial strain and, as we shall see, is a unique condition that uncouples the displacement and pressure variables. We will let x_3 be the direction of nonzero strain. Thus, p and u_3 are independent of x_1 and x_2 and $u_1 = u_2 = 0$. Given $E_{11} = E_{22} = 0$ and $E_{33} = \varepsilon$, Eq. (8.6) yields

$$T_{33} = (\lambda + 2\mu)\,\varepsilon - \alpha p \tag{8.21}$$

where we have dropped the subscript 'solid' from μ. Turning to Eq. (8.17) and assuming no source term ($Q = 0$),

$$S_E \frac{\partial p}{\partial t} - k \frac{\partial^2 p}{\partial x_3^2} = -\alpha \frac{\partial E_{kk}}{\partial t} = -\alpha \frac{\partial \varepsilon}{\partial t} \tag{8.22}$$

From Eq. (8.21),

$$\frac{\partial \varepsilon}{\partial t} = \frac{1}{(\lambda + 2\mu)} \frac{\partial T_{33}}{\partial t} + \frac{\alpha}{(\lambda + 2\mu)} \frac{\partial p}{\partial t} \tag{8.23}$$

Substituting Eq. (8.23) into Eq. (8.22) and rearranging yields

$$S_E \frac{\partial p}{\partial t} - k \frac{\partial^2 p}{\partial x_3^2} = -\frac{\alpha}{(\lambda + 2\mu)} \frac{\partial T_{33}}{\partial t} - \frac{\alpha^2}{(\lambda + 2\mu)} \frac{\partial p}{\partial t}$$

or (8.24)

$$\frac{\partial p}{\partial t} - k \left[S_E + \frac{\alpha^2}{(\lambda + 2\mu)} \right]^{-1} \frac{\partial^2 p}{\partial x_3^2} = -\frac{\alpha}{(\lambda + 2\mu)} \left[S_E + \frac{\alpha^2}{(\lambda + 2\mu)} \right]^{-1} \frac{\partial T_{33}}{\partial t}$$

Now, assuming no body forces and quasi-static loading, Eq. (3.21) for $i = 3$ becomes

$$\frac{\partial T_{31}}{\partial x_1} + \frac{\partial T_{32}}{\partial x_2} + \frac{\partial T_{33}}{\partial x_3} = \frac{\partial T_{33}}{\partial x_3} = 0 \tag{8.25}$$

which shows that T_{33} is independent of x_3 in this problem. Thus, we can replace the partial derivative of T_{33} with respect to time in Eq. (8.24) with the total time derivative. Thus, Eq. (8.24) can be written succinctly as

$$\frac{\partial p}{\partial t} - C_1 \frac{\partial^2 p}{\partial x_3^2} = -C_2 \frac{\mathrm{d}T_{33}}{\mathrm{d}t} \tag{8.26}$$

where we have collected the material coefficients into the constants C_1 and C_2.

Eq. (8.26) is a general result for the uniaxial consolidation problem. Let us consider the specific case when a step load is applied to the loading platen at $t = 0$, i.e., $T_{33} = -(F/A)H(t)$. For all time greater than $t = 0^+$, $\frac{dT_{33}}{dt} = 0$. Thus, Eq. (8.26) becomes

$$\frac{\partial p}{\partial t} = C_1 \frac{\partial^2 p}{\partial x_3^2} \tag{8.27}$$

The initial condition is

$$p\left(x_3, 0^+\right) = p_0 \tag{8.28}$$

and the boundary conditions are

$$p(0,t) = 0$$

and

$$\frac{\partial p}{\partial x_3}(L,t) = 0 \tag{8.29}$$

The first boundary condition reflects the fact that the platen is porous (i.e., free draining), and the second reflects the fact the permeability of the base is infinite (i.e., no flow; there is a zero pressure gradient). The solution to Eq. (8.27) subject to Eqs. (8.28) and (8.29) is

$$p(x_3,t) = 2p_0 \sum_{n=1}^{\infty} \frac{1}{\lambda_n} \sin\left(\lambda_n \frac{x_3}{L}\right) e^{-\lambda_n^2 \frac{C_1}{L^2} t} \tag{8.30}$$

where $\lambda_n = \frac{(2n-1)\pi}{2}$.

What about the displacement? Eqs. (8.25) and (8.21) imply

$$(\lambda + 2\mu) \frac{\partial^2 u_3}{\partial x_3^2} = \alpha \frac{\partial p}{\partial x_3} \tag{8.31}$$

Taking the derivative of Eq. (8.30) with respect to x_3, substituting the result into Eq. (8.31), and integrating twice with respect to x_3 yields

$$u_3(x_3,t) = -2p_0 L \left(\frac{\alpha}{\lambda + 2\mu}\right) \sum_{n=1}^{\infty} \frac{1}{\lambda_n^2} \cos\left(\lambda_n \frac{x_3}{L}\right) e^{-\lambda_n^2 \frac{C_1}{L^2} t} + f(t) x_3 + g(t) \tag{8.32}$$

Thus, we need two boundary conditions to complete the solution. The first boundary condition is $u_3(L,t) = 0$. This results in $g(t) = -f(t)L$. The second boundary condition comes from the stress condition and Eq. (8.21). We have

$$T_{33}(0,t) = \frac{-F}{A} = (\lambda + 2\mu)\frac{\partial u_3}{\partial x_3}(0,t) - \alpha p(0,t) \Rightarrow \frac{\partial u_3}{\partial x_3}(0,t) = \frac{-F}{A(\lambda + 2\mu)}$$

Applying this condition yields $f(t) = \frac{-F}{A(\lambda + 2\mu)}$. Thus, Eq. (8.32) becomes

$$u_3(x_3,t) = \frac{F}{A(\lambda + 2\mu)}(L - x_3) - 2p_0 L\left(\frac{\alpha}{\lambda + 2\mu}\right)\sum_{n=1}^{\infty}\frac{1}{\lambda_n^2}\cos\left(\lambda_n \frac{x_3}{L}\right)e^{-\lambda_n^2 \frac{C_1}{L^2}t} \tag{8.33}$$

For an incompressible fluid and solid, $\alpha = 1$ and $S_E = 0$ so that $C_1 = k(\lambda + 2\mu)$. Also under these assumptions, $p_0 = F/A$, as the initial stress is transferred immediately throughout the pore fluid. When the solid and fluid are considered incompressible, this problem is equivalent to confined compression of a biphasic material (see Chapter 9). For more about the equivalence between *incompressible* poroelasticity and the biphasic theory, see reference [25].

8.3 THERMOELASTICITY

8.3.1 Introduction and Fourier's Law

Thermoelasticity is concerned with how the temporal deformation of a body can induce temperature changes and vice versa. We present thermoelasticity in this chapter due to its mathematical equivalence to poroelasticity [24,26]. Thus, by becoming competent in the mathematics and principles of one theory, one becomes familiar with the other theory. However, note that the interpretation of the two theories is *not* equivalent. As we shall see shortly, stress in a thermoelastic material is borne, in part, by changes in a temperature field. For a poroelastic material, it is easy to visualize how the fluid can share the load. However, there is not as concrete a visualization for a temperature field. Without going too deep into the mechanisms behind thermoelastic coupling of stress and temperature, we offer the following greatly simplified explanation. Elementary physics allows us to idealize a solid as a lattice of atoms bonded together vibrating about an equilibrium position. Increased temperature increases the interatomic separation, thereby increasing the material's size (i.e., inducing strain). This is the coupling of temperature and deformation (i.e., thermal strain), which, when a body is confined in some way (i.e., not able to freely expand), induces thermal stresses. Another fundamental difference between the two theories is that the temperature field can support a shear stress.

Fourier's law is the Darcy equivalent in thermoelasticity. Empirically we know that 1. heat does not "flow" under conditions of constant temperature, 2. heat flows from hot to cold, 3. the greater the difference in temperature, the greater the heat flow, and 4. heat flow depends on the material through which it is flowing. These observations were summarized mathematically (and demonstrated experimentally) by Fourier as

$$\vec{q}_{\text{heat}} = -\underset{\approx}{\kappa}\nabla\Theta = -\underset{\approx}{\kappa}\nabla\theta \tag{8.34}$$

where q is the heat flux vector (heat per time per unit area, i.e., $\frac{\text{heat/time}}{\text{area}} = \frac{\text{heat}}{\text{time} \times \text{area}}$), and $\underset{\sim}{\kappa}$ is the anisotropic thermal conductivity tensor. Θ and θ are defined below. The negative sign ensures heat "flows" from hot to cold.

8.3.2 Constitutive and Governing equations ("u–θ" Formulation)

As it was for poroelasticity, there are several ways to formulate the constitutive equations of thermoelasticity depending on the choice of independent variables one desires to use. We will present the form that clarifies the analogy between the two theories. For further reading on thermoelasticity, see the references [27,28].

For stress in a linear anisotropic thermoelastic continuum,

$$T_{ij} = C_{ijkl}E_{kl} - b_{ij}\theta \tag{8.35}$$

If we consider an isotropic material with $\underset{\sim}{b} = \gamma \underset{\sim}{I}$, Eq. (8.35) becomes

$$T_{ij} = 2\mu E_{ij} + \lambda E_{kk}\delta_{ij} - \gamma\theta\delta_{ij} \tag{8.36}$$

In thermoelasticity, the analogous concept to increment of fluid content, is entropy, S. For isotropic materials

$$S = \gamma E_{kk} + \frac{c_\varepsilon}{\theta_0}\theta \tag{8.37}$$

where θ_0 is a constant reference temperature corresponding to the resting state of the material, i.e., undeformed and stress free. The variable θ is the deviation in the absolute temperature from the reference temperature, i.e., $\theta = \Theta - \theta_0$. Furthermore, these equations refer to conditions where θ does not deviate far from the reference temperature, i.e., $\left|\frac{\theta}{\theta_0}\right| \ll 1$.

$$\gamma \equiv \left.\frac{\partial S}{\partial E_{kk}}\right|_{\theta=\text{constant}} = (3\lambda + 2\mu)\,\alpha = 3K\alpha \tag{8.38}$$

The material coefficient γ relates the change in entropy to the change in volume at constant temperature. The coefficient α is the coefficient of linear thermal expansion (see problem 5), and K is the bulk modulus of the solid (see chapter 4). The material coefficient c_ε is the specific heat at constant strain. Recall that the specific heat of a substance is the quantity of heat required to raise the temperature of a unit mass of a substance by one degree. It is related to the "normal" specific heat by $c_\varepsilon = \rho c$. There are other constants that can be described for thermoelasticity. Importantly, many of the constants have statistical mechanical foundations and interpretations.

Now, for stress equilibrium, Eqs. (3.21) and (8.36) give

$$\frac{\partial}{\partial x_j}\left(2\mu E_{ij} + \lambda E_{kk}\delta_{ij} - \gamma\theta\delta_{ij}\right) + \rho B_i = \rho a_i \tag{8.39}$$

Taking the derivative, we obtain

$$\mu\frac{\partial^2 u_i}{\partial x_k^2} + (\lambda+\mu)\frac{\partial^2 u_k}{\partial x_i \partial x_k} = \left(\gamma\frac{\partial\theta}{\partial x_i} - \rho B_i\right) + \rho\frac{\partial^2 u_i}{\partial t^2} \tag{8.40}$$

This expression shows the coupling between displacement and temperature.

Now, we will examine the heat (h) balance. Heuristically, the time rate of change in heat is equal to any heat source less the divergence of heat. Reference [12] contains a good discussion of heat balance and the heat equation. Mathematically,

$$\frac{\partial h}{\partial t} = -\nabla\cdot\vec{q}_{\text{heat}} + Q(\vec{x},t) \tag{8.41}$$

We now recall the thermodynamic relation

$$\mathrm{d}S = \frac{\mathrm{d}h}{\theta_0} \Rightarrow h = S\theta_0 \tag{8.42}$$

Eq. (8.42) into Eq. (8.41) gives

$$\theta_0\frac{\partial S}{\partial t} = -\nabla\cdot\vec{q}_{\text{heat}} + Q(\vec{x},t) \tag{8.43}$$

Substituting Eqs. (8.34) and Eq. (8.37) with constant material coefficients into Eq. (8.43) results in

$$\theta_0\left(\gamma\frac{\partial E_{kk}}{\partial t} + \frac{c_\varepsilon}{\theta_0}\frac{\partial\theta}{\partial t}\right) = \kappa\nabla^2\theta + Q \tag{8.44}$$

Rearranging,

$$\left(\eta\frac{\partial E_{kk}}{\partial t} - \frac{Q}{\kappa}\right) = \nabla^2\theta - \frac{c_\varepsilon}{\kappa}\frac{\partial\theta}{\partial t} \tag{8.45}$$

where $\eta = \frac{\gamma\theta_0}{\kappa}$. Eq. (8.45) is the Heat equation with a source term. Here, the source term consists of everything within parentheses.

This completes the "u–θ" formulation of thermoelasticity. Eqs. (8.40) and (8.45) represent four equations for the four unknowns, u_1, u_2, u_3, and θ. They form a system of coupled PDEs that must be solved simultaneously subject to appropriate initial and boundary conditions on the displacements and temperature.

8.3.3 Thermal Prestress/Prestrain

If the equations describing the displacement and temperature can be uncoupled, then we have a special situation of thermal prestress/prestrain. We will look at two ways this may arise. The first is steady-state problems (i.e., all quantities are independent of time). In that case, Eqs. (8.40) and (8.45) reduce to

$$\mu \frac{\partial^2 u_i}{\partial x_k^2} + (\lambda + \mu) \frac{\partial^2 u_k}{\partial x_i \partial x_k} + \rho B_i = \gamma \frac{\partial \theta}{\partial x_i} \tag{8.46}$$

and

$$\nabla^2 \theta = -\frac{Q}{\kappa} \tag{8.47}$$

Eq. (8.47) is Poisson's equation in θ. We see that θ can be solved for independently of the equations of motion. After solving Eq. (8.47), the solution can then be used to solve Eq. (8.46) for the displacements (where it now acts as a known term). From the displacements, the strains can be calculated. Finally, stresses are calculated from the constitutive equation, Eq. (8.36). Steady-state cases are known as the stationary problems of thermoelasticity [27].

The second case leading to uncoupled equations occurs when we neglect the coupling term (equivalent to $\eta = 0$, i.e., η small, or an incompressible material, i.e., $E_{kk} = 0$). In either way, the equation for temperature becomes

$$\nabla^2 \theta - \frac{c_\varepsilon}{\kappa} \frac{\partial \theta}{\partial t} = -\frac{Q}{\kappa} \tag{8.48}$$

Eqs. (8.40) and (8.48) constitute the governing equations for the theory of thermal stresses [27]. As with stationary problems, we solve Eq. (8.48) for $\theta(\vec{x}, t)$ and use the result in Eq. (8.40) to obtain the displacements.

Demonstration. A rod of length L is held fixed at its ends in an undeformed, unstressed state. The lateral surface of the rod is insulated and rigidly confined. The rod, initially at a uniform temperature θ_0, is brought to a new temperature $\theta_c > \theta_0$ (by contact with thermal baths of temperature θ_c placed at each end of the rod), and steady-state conditions are allowed to be reached. The temperature increase is an example of a "thermal load." What is the axial stress generated in the rod during this process?

Solution. The problem described is stationary. Thus, we will solve for θ first, and then for the axial stress, T_{11}. For the 1-D case with no sources, Eq. (8.47) is

$$\frac{d^2 \theta}{dx_1^2} = 0$$

The solution to the above equation when $\theta(0) = \theta(L) = \theta_c$ is trivially found to be $\theta = \theta_c$. As the rod is held fixed at its ends and laterally confined, there is no deformation, i.e., $E_{11} = E_{22} = E_{33} = 0$. Substituting these results into Eq. (8.36) yields

$$T_{11} = -3K\alpha(\theta_c - \theta_0)$$

The negative sign indicates the rod is in a "state of compression" even though there has been no deformation.

In thermoelasticity, stress and temperature are state variables. Thus, though loading and temperature change may be occurring simultaneously, the problems of thermal prestress/strain can be analyzed in two parts: 1. determine the temperature distribution within the material, and 2. use the temperature distribution in the equations of motion to find the resulting displacements, strains, and stresses corresponding to the loads and temperature changes.

Finally, one may be wondering if there is such a thing as porothermoelasticity. The answer is a definite yes. There are also theories of poroviscoelasticity, thermoviscoelasticity, porohyperelasticity, thermohyperelasticity, and even poro-thermo-visco-hyper-elasticity! Of course, these all neglect the interaction of electric forces with the material, for which we have piezoelasticity and derivatives thereof. Finally, for deformations beyond the elastic limit, there are the complicated theories of plasticity. One can see that a wealth of theories has been developed for describing material behavior.

8.4 PROBLEMS

1. In this chapter, we introduced the permeability tensor and thermal conductivity tensor. Second-order tensors obey the following transformation law

$$T'_{ij} = Q_{ri} Q_{sj} T_{rs} \tag{8.49}$$

   In the manner of Section 4.11, express $\underset{\sim}{\kappa}$
   (a) for a material orthotropic with respect to permeability
   (b) for a material transversely isotropic with respect to permeability
2. Under quasi-static conditions in the absence of body forces and fluid sources, show that the governing equation for the dilatation of a poroelastic medium consisting of an incompressible solid matrix and infinitely incompressible fluid is given by

$$\frac{\partial E_{kk}}{\partial t} = k(\lambda + 2\mu_{\text{solid}})\, \nabla^2 E_{kk} \tag{8.50}$$

   Eq (8.50) equation resembles the "Heat" equation for E_{kk}.

3. In the absence of body forces, we derived Eq. (8.26) as the governing equation for pressure during a uniaxial consolidation experiment. Solve Eq. (8.26) for incompressible solid and fluid components when the loading rate is constant, i.e., $T_{33} = -\sigma t$.
4. Write Fourier's law for a material orthotropic with respect to thermal conductivity. What is $-\nabla \cdot \vec{q}_{\text{heat}}$ for a homogeneous material? For an inhomogeneous material?
5. For unconstrained thermal expansion, no stress is produced, i.e., $\underset{\sim}{T} = 0$.
   (a) Use Eq. (8.36) to show that the dilatation during unconstrained thermal expansion is $E_{kk} = \dfrac{\gamma}{K}\theta$.
   (b) Using Eq. (8.38), show that the change in volume per unit volume is $3\alpha\theta$, where α is the coefficient of linear thermal expansion (i.e., $E_{11} = \alpha\theta$).
   (c) A material initially has a volume of 1 m^3. After being heated from 25°C to 3,025°C, the volume is 3 m^3. What is the coefficient of linear thermal expansion? You may assume that the coefficient of thermal expansion is constant over this temperature range.
6. An undeformed, unstressed rod of temperature θ_0 and length L is fixed at $x_1 = 0$. A thermal bath of temperature θ_1 is brought into contact with the bar at $x_1 = 0$. At $x_1 = L$, the rod is brought into contact with a thermal bath of temperature θ_2. Furthermore, the bath at $x_1 = L$ is pulled in the positive x_1 direction, creating a state of uniaxial tension (see Chapter 4). What is the axial strain in the rod at steady state? You may assume the material is isotropic and homogeneous. Also, assume no body forces, no additional heat sources, a traction-free lateral surface, and quasi-static, infinitesimal loading.
7. Consider when we neglect the coupling term, η. Assuming no heat sources, repeat the last demonstration in this chapter to find the transient axial stress, i.e., $T_{11}(x_1,t)$, when the initial temperature is $\theta(x_1,0) = \theta_0$. Show that, in the limit as $t \to \infty$, $T_{11} = -3K\alpha(\theta_c - \theta_0)$. Is the initial stress greater or less than the equilibrium stress? What does this mean for the design of materials that may experience thermal loads?
8. The solution for the shear stress in an infinite thermoelastic medium when a point source of heat equal to h is supplied to the origin at time $t = 0$ is [28]

$$T_{12} = \frac{3}{R^4}\left(\frac{\gamma\mu h}{2\pi c_\varepsilon(\lambda + 2\mu)}\right)(x_1 x_2)\left[\frac{1}{\sqrt{\pi(\kappa/c_\varepsilon)t}}\left(1 + \frac{R^2}{6(\kappa/c_\varepsilon)t}\right)e^{\frac{-R^2}{4(\kappa/c_\varepsilon)t}} - \frac{1}{R}\mathrm{erf}\left(\frac{R}{2\sqrt{(\kappa/c_\varepsilon)t}}\right)\right]$$

where $\mathrm{erf}(x) = \dfrac{2}{\sqrt{\pi}}\displaystyle\int_0^x e^{-x^2}dx$.

An infinite poroelastic medium has an amount of fluid V_f added to the origin at time $t = 0$. What is T_{12}?

• • • •

CHAPTER 9

Biphasic Theory

9.1 INTRODUCTION

Biphasic theory is a simplification of mixture theory, where there are only two phases or constituents. In general, mixture theory describes a material as a continuum mixture of n phases, such that each spatial point in the mixture is occupied simultaneously by all the phases. This aspect of *superimposed continua* is extremely important. It allows mathematical simplification, as we do not have to follow individual spatial points corresponding to separate phases to describe the tissue's mechanical behavior. Each of the phases has a density and displacement field, and thus, each also has continuity, momentum, and energy equations. Truesdell constructed three guiding principles for the theory of mixtures as follows: 1) every property of the mean motion of the mixture is a mathematical consequence of the properties of the motion of the constituents, 2) the balance laws for the mixture as a whole have the same form as the equations for a single phase mixture, and 3) if one begins with n phases, and the volume fraction of one of those phases is equal to zero, then the equations should reduce to those for a material composed of $n-1$ phases. From these principles, it can be concluded that separate conservation equations can be constructed for each constituent and then added together to obtain the conservation equation for the entire mixture [29,30]. However, this seemingly simple operation is complicated by the fact that new terms must be introduced to each constituent's constitutive equation to describe the interaction (exchange of mass or momentum) between the particular constituent and all other mixture constituents, a fact we will describe later.

In this chapter our objective will be to describe the rheological behavior of materials made of binary mixtures. Instead of using the exact geometry of the microstructure, equations will be developed to describe the macroscopic structure using mixture theory. The biphasic theory has been identified as a suitable framework to describe the constitutive behavior of soft, hydrated biological materials such as articular cartilage, fibrocartilages (e.g., the meniscus, temporomandibular joint disc, and intervertebral disc), and even cells. It has also been applied in tissue engineering. Common to these tissues or tissue-engineered constructs is viscoelastic behavior that results from the flow of fluid through a solid network containing pores of small size. This is known as flow-dependent viscoelasticity, as the solid matrix may, by itself, have intrinsic flow-independent viscoelasticity. These

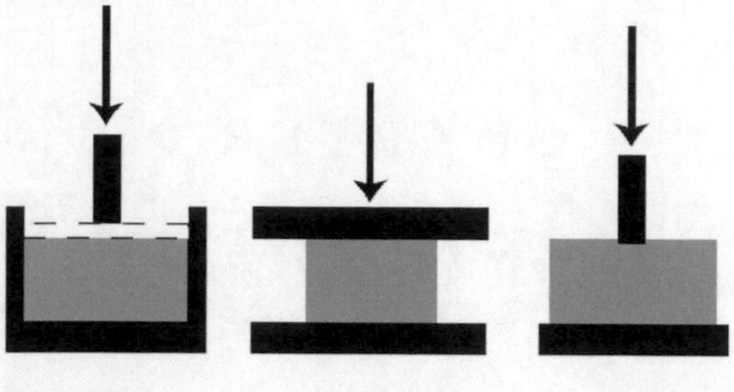

FIGURE 9.1: Illustration of experimental setup for confined, unconfined, and indentation compressive testing. See text for description.

materials also have small permeabilities, slowing the flow of fluid through the solid network and allowing an interstitial fluid pressure to be generated.

There are three common compressive testing modalities to which the biphasic theory has been applied: 1) confined compression, 2) unconfined compression, and 3) indentation (see Figure 9.1). Furthermore, each of these modalities can be used for creep or stress relaxation. Let us briefly discuss these three testing setups. Later in the chapter, we will examine some of the mathematics used to model confined and unconfined compression. *Confined compression* is an idealized deformation configuration in which a tissue specimen is placed into a chamber that confines the specimen at the bottom and on the sides. A porous platen, which allows fluid to exit through the specimen's surface, is used to compress the sample. *Unconfined compression*, most easily discussed in cylindrical coordinates, is a test setup where a tissue specimen is confined between two rigid, nonporous, frictionless platens, but is free to expand in the radial direction. An axial load or deformation is applied, and the radial deformation is monitored. Finally, *indentation*, another testing setup most conveniently described in cylindrical coordinates, is where the center of a tissue specimen is indented by a frictionless cylindrical probe that can be either porous or non-permeable.

9.2 DEFINITIONS

Consider an infinitesimal volume, $\mathrm{d}V$, consisting of a fluid of volume $\mathrm{d}V^{\mathrm{f}}$ and mass $\mathrm{d}m^{\mathrm{f}}$, and a solid of volume $\mathrm{d}V^{\mathrm{s}}$ and mass $\mathrm{d}m^{\mathrm{s}}$. From these masses and volumes, two densities can be defined. The true density of each constituent is

$$\rho_T^{\mathrm{f}} = \frac{\mathrm{d}m^{\mathrm{f}}}{\mathrm{d}V^{\mathrm{f}}} \quad \text{and} \quad \rho_T^{\mathrm{s}} = \frac{\mathrm{d}m^{\mathrm{s}}}{\mathrm{d}V^{\mathrm{s}}} \tag{9.1}$$

The apparent density is

$$\rho^{\mathrm{f}} = \frac{\mathrm{d}m^{\mathrm{f}}}{\mathrm{d}V} \quad \text{and} \quad \rho^{\mathrm{s}} = \frac{\mathrm{d}m^{\mathrm{s}}}{\mathrm{d}V} \tag{9.2}$$

From Eqs. (9.1) and (9.2), *volume fractions* of the fluid and solid can be defined as

$$\phi^{\mathrm{f}} = \frac{\mathrm{d}V^{\mathrm{f}}}{\mathrm{d}V} = \frac{\rho^{\mathrm{f}}}{\rho_T^{\mathrm{f}}} \quad \text{and} \quad \phi^{\mathrm{s}} = \frac{\mathrm{d}V^{\mathrm{s}}}{\mathrm{d}V} = \frac{\rho^{\mathrm{s}}}{\rho_T^{\mathrm{s}}} \tag{9.3}$$

ϕ^{f} is also known as the *porosity* of the material, and ϕ^{s} is known as the *solidity*. Note that, given n constituents in the mixture, the following must hold

$$\sum_{i=1}^{n} \phi^{i} = 1 \tag{9.4}$$

This equation constrains the mass to remain within the mixture. If, for some reason, ϕ^{i} of one constituent was to change, e.g., decrease, the other $n-1$ ϕ^{i}'s would have to change, e.g., increase, to keep Eq. (9.4) true.

9.3 CONSERVATION OF MASS

Using Eq. (2.40), the continuity equation for the fluid and solid becomes

$$\rho^{\mathrm{f}} \left(\operatorname{div} \vec{v}^{\mathrm{f}}\right) + \frac{\mathrm{D}\rho^{\mathrm{f}}}{\mathrm{D}t} = 0 \tag{9.5}$$

and

$$\rho^{\mathrm{s}} \left(\operatorname{div} \vec{v}^{\mathrm{s}}\right) + \frac{\mathrm{D}\rho^{\mathrm{s}}}{\mathrm{D}t} = 0 \tag{9.6}$$

Assuming the true densities of the fluid and solid are constant (i.e., the fluid and solid are intrinsically incompressible), substitution of Eq. (9.3) into Eqs. (9.5) and (9.6) yields

$$\frac{\partial \phi^{\mathrm{f}}}{\partial t} + \nabla \cdot \left(\phi^{\mathrm{f}} \vec{v}^{\mathrm{f}}\right) = 0 \tag{9.7}$$

and

$$\frac{\partial \phi^{s}}{\partial t}+\nabla\cdot\left(\phi^{s}\vec{v}^{s}\right)=0 \tag{9.8}$$

Adding Eqs. (9.7) and (9.8), and using Eq. (9.4), we arrive at the continuity equation for a biphasic system containing incompressible fluid and solid constituents

$$\nabla\cdot\left(\phi^{f}\vec{v}^{f}+\phi^{s}\vec{v}^{s}\right)=0 \tag{9.9}$$

Demonstration. Consider a confined compression setup where a displacement of magnitude $u_3(t)$, is imposed on the top face of the tissue by a rigid, impermeable platen of cross-sectional area, A (see Figure 9.1). At the bottom face of the tissue there is an immobile, rigid, porous platen also of area A. Determine the volumetric fluid flux through the bottom platen.
Solution. The confined compression set-up is a 1-D deformational configuration. In long form, Eq. (9.9) is

$$\nabla\cdot\left[\phi^{f}\left(v_1^{f}\vec{e}_1+v_2^{f}\vec{e}_2+v_3^{f}\vec{e}_3\right)+\phi^{s}\left(v_1^{s}\vec{e}_1+v_2^{s}\vec{e}_2+v_3^{s}\vec{e}_3\right)\right]=0$$
$$\nabla\cdot\left(\phi^{f}v_3^{f}\vec{e}_3+\phi^{s}v_3^{s}\vec{e}_3\right)=\nabla\cdot\left(\phi^{f}v_3^{f}+\phi^{s}v_3^{s}\right)\vec{e}_3=0$$

where the second step follows, as the only direction that matters is the direction coincident with the axis of loading, x_3 in this case. Performing the dot product, we get

$$\frac{\partial}{\partial x_3}\left(\phi^{f}v_3^{f}+\phi^{s}v_3^{s}\right)=0$$

Integrating with respect to x_3,

$$\phi^{f}v_3^{f}+\phi^{s}v_3^{s}=\text{constant}$$

This implies that

$$\left.\left(\phi^{f}v_3^{f}+\phi^{s}v_3^{s}\right)\right|_{\text{Top}}=\left.\left(\phi^{f}v_3^{f}+\phi^{s}v_3^{s}\right)\right|_{\text{Bottom}} \tag{9.10}$$

Now, at the top surface, $v_3^{f}=v_3^{s}=\dfrac{\partial u_3}{\partial t}$, and at the bottom surface, $v_3^{s}=0$. At the bottom surface, v_3^{f} is equal to the volumetric flow rate divided by the cross-sectional area of the surface available for permeation, $v_3^{\mathrm{f}} = \dfrac{Q}{A\phi^{\mathrm{f}}}$. Note that Darcy's law for fluid flux through a porous medium describes relative fluid velocity, v, in terms of total area, $v = Q/A$, whereas the biphasic theory describes fluid velocity in terms of the actual permeation area, $v^{\mathrm{f}} = v/\phi^{\mathrm{f}}$. Substituting these values into Eq. (9.10) gives the final answer,

$$Q = A\frac{\partial u_3}{\partial t}$$

9.4 CONSERVATION OF MOMENTUM

From Eq. (3.21), we can write for the fluid and solid phases of the mixture

$$\frac{\partial T_{ij}^{\mathrm{f}}}{\partial x_j} + \left(\rho^{\mathrm{f}} B_i^{\mathrm{f}} + \Pi_i^{\mathrm{f}}\right) = \rho^{\mathrm{f}} a_i^{\mathrm{f}} \tag{9.11}$$

and

$$\frac{\partial T_{ij}^{\mathrm{s}}}{\partial x_j} + \left(\rho^{\mathrm{s}} B_i^{\mathrm{s}} + \Pi_i^{\mathrm{s}}\right) = \rho^{\mathrm{s}} a_i^{\mathrm{s}} \tag{9.12}$$

The term $(\rho^{\mathrm{f}} B_i^{\mathrm{f}} + \Pi_i^{\mathrm{f}})$ is worth discussing. In this term, B_i^{f} represents the sum of external body forces per unit mass acting on the fluid, and Π_i^{f} is the term describing internal body forces reflecting the interaction between the fluid and solid. The term $(\rho^{\mathrm{f}} B_i^{\mathrm{f}} + \Pi_i^{\mathrm{f}})$ reflects non-inertial and non-deformation-induced forces. The term in the equation for the solid phase is interpreted similarly. Summing Eqs. (9.11) and (9.12), we get

$$\frac{\partial}{\partial x_j}\left(T_{ij}^{\mathrm{f}} + T_{ij}^{\mathrm{s}}\right) + \left(\rho^{\mathrm{f}} B_i^{\mathrm{f}} + \rho^{\mathrm{s}} B_i^{\mathrm{s}}\right) + \left(\Pi^{\mathrm{f}} + \Pi^{\mathrm{s}}\right) = \left(\rho^{\mathrm{f}} a_i^{\mathrm{f}} + \rho^{\mathrm{s}} a_i^{\mathrm{s}}\right) \tag{9.13}$$

Noting Truesdell's second principle, and comparing Eq. (9.13) to Eq. (3.21), we identify that

$$\begin{aligned} \rho B_i &= \left(\rho^{\mathrm{f}} B_i^{\mathrm{f}} + \rho^{\mathrm{s}} B_i^{\mathrm{s}}\right) + \left(\Pi_i^{\mathrm{f}} + \Pi_i^{\mathrm{s}}\right) \\ \rho a_i &= \left(\rho^{\mathrm{f}} a_i^{\mathrm{f}} + \rho^{\mathrm{s}} a_i^{\mathrm{s}}\right) \end{aligned} \tag{9.14}$$

From Eq. (9.14), we see the total acting body force is the density weighted sum of the external body forces acting on the individual phases and the internal body force terms. We also see that the overall acceleration within the body is a density weighted sum of the acceleration of the solid and fluid phases. Let us now assume quasi-static equilibrium and no external body forces acting on the individual phases. This leaves only Π^f_i and Π^s_i, which we will assume are nonzero. Then, by Newton's third law, $\Pi^f_i = -\Pi^s_i$, resulting in zero net total acting body force. So, for a mixture of n constituents, it is required that

$$\sum_{i=1}^{n} \vec{\Pi}^i = 0 \tag{9.15}$$

Finally, simplifying Eqs. (9.11) and (9.12) under quasi-static conditions with no external body forces yields

$$\frac{\partial T^{\mathrm{f}}_{ij}}{\partial x_j} + \Pi^{\mathrm{f}}_i = 0 \tag{9.16}$$

and

$$\frac{\partial T^{\mathrm{s}}_{ij}}{\partial x_j} + \Pi^{\mathrm{s}}_i = 0 \tag{9.17}$$

Note that the sum of Eqs. (9.16) and (9.17), subject to the restriction $\Pi^f_i = -\Pi^s_i$, is $\nabla \cdot \underset{\sim}{T} = 0$, in accordance with Truesdell's second principle.

In general, we are free to choose whatever we would like for Π^f_i or Π^s_i and T^f_{ij} and T^s_{ij}. For example, the interaction body force could be defined as a cubic function of the current strain (i.e., $\Pi^s_i = \alpha_j E^3_{ij} + \beta_j E^2_{ij} + \chi_j E_{ij} + \delta_i$) or be a constant that depends on the temperature (i.e., $\Pi^s_i = C_i(\theta)$). Also, the stress tensors describing the fluid and solid could be any of the constitutive equations we have already discussed (i.e., hyperelastic, linearly elastic, Newtonian fluid, viscoelastic, thermoelastic, etc.) and many more. Of course, the choices made for both the internal body forces and constitutive equations should be based on experimental data, the level of sophistication required to "accurately" model the problem of interest (as more complicated choices can lead to intractable mathematics and numerical simulations) and, of course, good old fashioned intuition.

However, for our discussion, which suffices for the majority of problems in soft tissue biomechanics, we will model the interaction between the fluid and solid as a viscous drag force proportional to the relative velocity between the two phases as was originally done by Mow and coworkers [31]. We will see that this simple assumption captures the gross rheological behavior of articular cartilage under various loading conditions. Thus, the equations for Π_i^{f} and Π_i^{s} are those used to describe a viscous damper. They are

$$\vec{\Pi}^{\mathrm{f}} = K\left(\vec{v}^{\mathrm{s}} - \vec{v}^{\mathrm{f}}\right)$$

and (9.18)

$$\vec{\Pi}^{\mathrm{s}} = -\vec{\Pi}^{\mathrm{f}} = K\left(\vec{v}^{\mathrm{f}} - \vec{v}^{\mathrm{s}}\right)$$

where K is the coefficient of diffusive resistance and has units of N·s/m. Please see problem 4 for how K relates to the permeability, k, of Darcy's law. With Eq. (9.18), the conservation of momentum equations for our biphasic mixture become

$$\frac{\partial T_{ij}^{\mathrm{f}}}{\partial x_j} + K\left(v_i^{\mathrm{s}} - v_i^{\mathrm{f}}\right) = 0$$

$$\frac{\partial T_{ij}^{\mathrm{s}}}{\partial x_j} + K\left(v_i^{\mathrm{f}} - v_i^{\mathrm{s}}\right) = 0 \qquad (9.19)$$

We will now turn to a discussion of the constitutive equations used in the classical linear biphasic theory [31].

9.5 CONSTITUTIVE EQUATIONS

The next choices we have to make are what constitutive equations will be used to relate the stresses in the fluid and solid phases to the hydrostatic pressure, strain, and strain rate generated in the tissue upon loading. In this text, we have simply quoted constitutive equations. This is because the development of constitutive equations is either an entirely empirical process or a complex process depending on principles such as the second law of thermodynamics, material reference frame indifference, etc. [1], which are beyond the scope of this book. Using the latter approach, and considering infinitesimal deformations, Mow and coworkers arrived at the following constitutive equations for the fluid and solid phases

$$T_{ij}^{\mathrm{f}} = -\phi^{\mathrm{f}} p\delta_{ij} + \lambda_f D_{kk}^{\mathrm{f}}\delta_{ij} + 2\mu_f D_{ij}^{\mathrm{f}} - QE_{kk}\delta_{ij} + 2K_C\Gamma_{ij} \qquad (9.20)$$

and

$$T_{ij}^{\mathrm{s}} = -\phi^{\mathrm{s}} p\delta_{ij} + \lambda E_{kk}\delta_{ij} + 2\mu E_{ij} + \lambda_s D_{kk}^{\mathrm{s}}\delta_{ij} + 2\mu_s D_{ij}^{\mathrm{s}} - 2K_C\Gamma_{ij}, \qquad (9.21)$$

respectively, where p is the hydrostatic pressure in the system, and is proportioned to the solid and fluid phases according to their volume fractions (i.e., the total hydrostatic pressure is $p = \phi^f p + \phi^s p$). Furthermore, δ_{ij} is the Kronecker delta, E_{ij} are components of the infinitesimal strain tensor, D_{ij} are components of the strain-rate tensor, and $\Gamma_{ij} = W^s_{ij} - W^f_{ij}$ (recall $\underset{\sim}{W}$ is the spin tensor). The remaining terms are material constants. Notice that the first three terms in $\underset{\sim}{T}^f$ represent an isotropic Newtonian fluid (Eq. (5.17)), and the first three terms in $\underset{\sim}{T}^s$ represent an isotropic, linearly elastic solid (Eq. (4.31)) with the contribution of hydrostatic pressure in the tissue. The remainder of $\underset{\sim}{T}^s$ represents the viscous nature of the solid matrix should it need to be included.

We shall now apply several simplifying assumptions to these constitutive equations. As is commonly, and was originally done, we will treat the fluid phase as inviscid and neglect the viscous nature of the solid matrix. Furthermore, let $Q = K_C = 0$. Eqs. (9.20) and (9.21) then become

$$T^f_{ij} = -\phi^f p \delta_{ij} \tag{9.22}$$

and

$$T^s_{ij} = -\phi^s p \delta_{ij} + \lambda E_{kk} \delta_{ij} + 2\mu E_{ij} \tag{9.23}$$

The fluid stress in classical biphasic theory is only due to an apparent hydrostatic pressure; there is no rate of strain or rate of dilatation-induced stress contributions. The solid stress is the sum of the apparent hydrostatic pressure, as well as dilatational- and strain-induced stresses. It is worth pointing out that the constants of the solid are apparent, or matrix, properties (i.e., properties of the solid continuum), and are not the intrinsic structural properties of the solid constituents. For example, in articular cartilage, these properties are the properties of the solid matrix, and not the properties of specific components (i.e., collagens and proteoglycans).

9.6 SUMMARY AND EQUATIONS OF MOTION

So far, we have made several assumptions in the development of the biphasic theory equations. These assumptions are the following:

1. at every point in the tissue, both the solid and fluid are present
2. the solid and fluid phases are intrinsically incompressible, and therefore $\rho_T{}^f$ and $\rho_T{}^s$ are constant (requiring the total tissue volume to remain constant)
3. the strains are infinitesimal, and therefore ϕ^f and ϕ^s are constant
4. the solid "matrix" is not incompressible, but isotropic and linearly elastic (think of a truss with rigid membranes attached by elastic hinges)
5. the fluid is inviscid
6. tissue permeability (i.e., the resistance to flow, or drag coefficient) is constant

7. inertial effects are negligible
8. viscoelastic (i.e., time dependent) effects are caused by frictional drag associated with fluid flow through the solid matrix (i.e., no flow-independent viscoelasticity)

Assuming that the fluid and solid are intrinsically incompressible, we derived the continuity equation for the biphasic mixture as

$$\nabla \cdot \left(\phi^{\mathrm{f}}\vec{v}^{\mathrm{f}} + \phi^{\mathrm{s}}\vec{v}^{\mathrm{s}}\right) = 0 \tag{9.24}$$

Furthermore, choosing to model the interaction force between the fluid and solid phases as a viscous drag force and assuming quasi-static conditions, we arrived at the following equations for the balance of momentum

$$\begin{aligned} \nabla \cdot \underset{\sim}{T}^{\mathrm{f}} + K\left(\vec{v}^{\mathrm{s}} - \vec{v}^{\mathrm{f}}\right) &= 0 \\ \nabla \cdot \underset{\sim}{T}^{\mathrm{s}} + K\left(\vec{v}^{\mathrm{f}} - \vec{v}^{\mathrm{s}}\right) &= 0 \end{aligned} \tag{9.25}$$

We then assumed infinitesimal deformations and chose an inviscid fluid phase and a linearly elastic, isotropic solid phase to describe the constituents, described by the following constitutive equations

$$\begin{aligned} \underset{\sim}{T}^{\mathrm{f}} &= -\phi^{\mathrm{f}} p \underset{\sim}{I} \\ \underset{\sim}{T}^{\mathrm{s}} &= -\phi^{\mathrm{s}} p \underset{\sim}{I} + \lambda E_{kk} \underset{\sim}{I} + 2\mu \underset{\sim}{E} \end{aligned} \tag{9.26}$$

Inserting Eqs. (9.26) into Eqs. (9.25), we get the equations for motion of the biphasic system. They are given as

$$-\nabla\left(\phi^{\mathrm{f}} p\right) + K\left(\vec{v}^{\mathrm{s}} - \vec{v}^{\mathrm{f}}\right) = 0 \tag{9.27}$$

and

$$-\nabla\left(\phi^{\mathrm{s}} p\right) + \lambda \nabla E_{kk} + 2\mu \nabla \cdot \underset{\sim}{E} + K\left(\vec{v}^{\mathrm{f}} - \vec{v}^{\mathrm{s}}\right) = 0 \tag{9.28}$$

Note that these two equations each represent three equations; one for each of the principal coordinate directions. For example, considering the x_3 direction, i.e., taking the dot product of Eqs. (9.27) and (9.28) with $\vec{e}_3$, the equations of motion are

$$-\phi^{\mathrm{f}} \frac{\partial p}{\partial x_3} + K\left(v_3^{\mathrm{s}} - v_3^{\mathrm{f}}\right) = 0 \tag{9.29}$$

$$-\phi^{\mathrm{s}}\frac{\partial p}{\partial x_3}+\lambda\frac{\partial}{\partial x_3}(E_{11}+E_{22}+E_{33})+2\mu\left(\frac{\partial E_{31}}{\partial x_1}+\frac{\partial E_{32}}{\partial x_2}+\frac{\partial E_{33}}{\partial x_3}\right)$$
$$+K\left(v_3^{\mathrm{f}}-v_3^{\mathrm{s}}\right)=0 \qquad (9.30)$$

Demonstration. Consider a state of uniform hydrostatic pressure, $T_{ij}=-p\delta_{ij}$. What are the stresses in the fluid and solid phase?

Solution. There is no pressure gradient, so $\nabla(\phi^{\mathrm{f}}p) = \nabla(\phi^{\mathrm{s}}p) = 0$, and no fluid flows into the tissue. Also, there is no relative motion between the two phases, as force is balanced in all directions. Hence, $\vec{v}^{\mathrm{f}} - \vec{v}^{\mathrm{s}} = 0$. Furthermore, as both phases are incompressible, the change in volume is zero, i.e., $E_{kk} = 0$. Thus, we are left with $2\mu\nabla\cdot\underset{\sim}{E} = 0$ from Eq. (9.28), which implies $\nabla\cdot\underset{\sim}{E} = 0$. But, because of incompressibility, $E = 0$ under a state of hydrostatic pressure. Hence, using Eqs. (9.26), $\underset{\sim}{T}^{\mathrm{f}} = -\phi^{\mathrm{f}}p\underset{\sim}{I}$ and $\underset{\sim}{T}^{\mathrm{s}} = -\phi^{\mathrm{s}}p\underset{\sim}{I}$.

We will now examine two more complicated loading conditions, confined and unconfined compression.

9.7 CONFINED COMPRESSION

The confined compression experiment is a 1-D loading situation because we assume lateral confinement prevents displacement and fluid flow in the radial direction (or, equivalently, x_1 and x_2 directions). Setting the x_3 direction as the direction of loading, the displacement of the solid matrix is a function of x_3 and t alone.

$$u_1 = u_2 = 0$$
$$u_3 = u_3(x_3,t)$$

Thus, the strain tensor is

$$[\underset{\sim}{E}] = \begin{bmatrix} 0 & 0 & 0 \\ 0 & 0 & 0 \\ 0 & 0 & \partial u_3/\partial x_3 \end{bmatrix} \qquad (9.31)$$

from which the dilatation, E_{kk}, is calculated to be $\dfrac{\partial u_3}{\partial x_3}$. It follows that $\nabla E_{kk} = \left(\dfrac{\partial^2 u_3}{\partial x_3^2}\right)\vec{e}_3$ and $\nabla\cdot E = \left(\dfrac{\partial^2 u_3}{\partial x_3^2}\right)\vec{e}_3$. Furthermore, the velocity of the solid matrix is $\vec{v}^{\,s} = \left(\dfrac{\partial u_3}{\partial t}\right)\vec{e}_3$. Using these relationships, Eqs. (9.29) and (9.30) can be combined (see problem 4) to yield

$$H_A k \frac{\partial^2 u_3}{\partial x_3^2} = \frac{\partial u_3}{\partial t} \tag{9.32}$$

which is the governing equation for the confined compression configuration. H_A is the *aggregate modulus* (units of stress, i.e., Pa), and k is the *Darcy permeability* (units $m^4/N{\cdot}s$). Eq. (9.32) is of the same form as the 1-D diffusion or heat transfer equation covered in a first course on PDEs. It can be solved by separation of variables with appropriate boundary and initial conditions [12].

9.7.1 Creep

As described in Chapter 7, creep is deformation of a body under an applied constant load. The initial condition is

$$u_3(x_3, 0) = 0 \tag{9.33}$$

The first boundary condition is at the bottom of the confinement chamber where there is no displacement

$$u_3(h, t) = 0 \tag{9.34}$$

The second boundary condition reflects the fact that the solid matrix at the tissue surface supports the applied load throughout the deformation. If we designate the applied load as F, then the stress on that surface is

$$T_{33}^s(0, t) = -\frac{F}{A} \tag{9.35}$$

where A is the surface area of the tissue (and loading platen). Using Eqs. (9.31) and (9.23), and noting that there is no hydrostatic pressure at the surface (for a free-draining platen),

$$T_{33}^s = \lambda \frac{\partial u_3}{\partial x_3} + 2\mu \frac{\partial u_3}{\partial x_3} = H_A \frac{\partial u_3}{\partial x_3} \tag{9.36}$$

so that

$$-\frac{F}{A} = H_A \frac{\partial u_3}{\partial x_3}(0, t) \quad \Rightarrow \quad \frac{\partial u_3}{\partial x_3}(0, t) = -\frac{F}{H_A A} \tag{9.37}$$

We have transformed the stress boundary condition into a displacement boundary condition. Eq. (9.32), with the conditions (9.33), (9.34), and (9.37), constitute a nonhomogeneous PDE. The solution is

$$u_3(x_3, t) = \frac{F}{H_A A}(h - x_3) + \sum_{n=1}^{\infty} c_n \cos\left(\lambda_n \frac{x_3}{h}\right) e^{-\lambda_n^2 \frac{H_A k}{h^2} t} \tag{9.38}$$

where

$$\lambda_n = \frac{(2n-1)\pi}{2}$$

$$c_n = \left(\frac{F}{H_A \mathrm{A}}\right)\left(\frac{2h}{\lambda_n^2}\right)(\cos(\lambda_n) - 1) = -\left(\frac{F}{H_A A}\right)\left(\frac{2h}{\lambda_n^2}\right) \tag{9.39}$$

Experimentally, the most easily accessible displacement that can be measured is at the tissue's surface. At $x_3 = 0$, Eq. (9.38) becomes

$$\begin{aligned} u_3(0,t) &= \frac{F}{H_A A}h + \sum_{n=1}^{\infty} c_n \mathrm{e}^{-\lambda_n^2 \frac{H_A k}{h^2} t} \\ &= \frac{F}{H_A A}h\left[1 - 2\sum_{n=1}^{\infty}\frac{1}{\lambda_n^2}\mathrm{e}^{-\lambda_n^2 \frac{H_A k}{h^2} t}\right] \end{aligned} \tag{9.40}$$

A plot of this solution using the first 100 terms of the sum is shown in Figure 9.2. Let us make a few observations concerning Eq. (9.40). First, each term in the series solution decays with a time constant equal to $h^2/H_A k$, known as the "gel diffusion time (t_D)." The "gel diffusion time" increases with decreasing H_A and k and increasing h. A typical value of the "gel diffusion time" for articular cartilage is ~1,000 s. The properties H_A and k can be determined by curve fitting experimental data; H_A is calculated by examining the curve at $t \to \infty$, at which time the fluid pressure and fluid flow have gone to zero, and the entire load is supported by elastic stresses in the solid matrix. The equation for finding H_A is

$$H_A = \frac{hF}{u_3(0,\infty)A} \tag{9.41}$$

Next, k can be determined by nonlinear curve fitting Eq. (9.40) with the calculated value of H_A and a measured value of the tissue's height. For example, one could import the creep data into Matlab© and use a nonlinear curve fitting algorithm (see problem 6). Rewriting Eq. (9.41), we get

$$\frac{u_3(0,\infty)}{h}H_A = \frac{F}{A} \quad \rightarrow \quad \sigma = H_A \varepsilon \tag{9.42}$$

where σ is the stress in the solid matrix, and ε is the bulk tissue strain, an expression reminiscent of Hooke's law. Let us now examine the pressure distribution in the tissue. An intermediate step of problem 4 shows that

$$\frac{\partial p}{\partial x_3} = \frac{1}{k}\frac{\partial u_3}{\partial t} \tag{9.43}$$

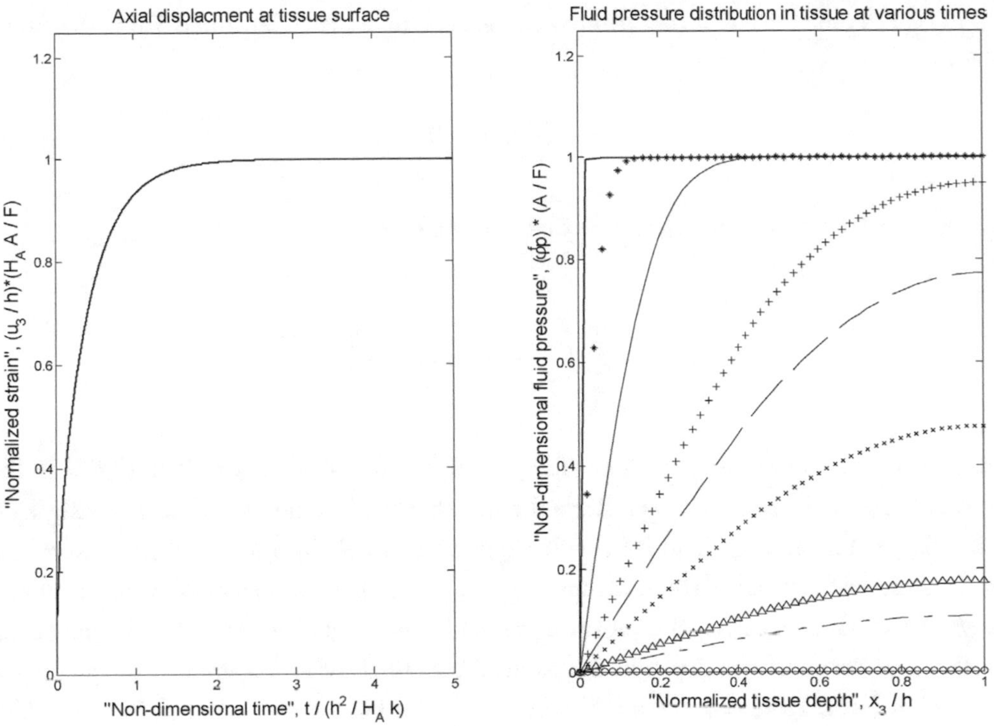

FIGURE 9.2: Strain and pressure resulting from a confined compression creep experiment.

Taking the time derivative of Eq. (9.38) term by term results in

$$
\begin{aligned}
\frac{\partial p}{\partial x_3} &= \frac{1}{k}\left[\sum_{n=1}^{\infty}\left(-\lambda_n^2\frac{H_A k}{h^2}\right) c_n \cos\left(\lambda_n\frac{x_3}{h}\right) \mathrm{e}^{-\lambda_n^2\frac{H_A k}{h^2}t}\right] \\
&= \sum_{n=1}^{\infty}\left(-\lambda_n^2\frac{H_A}{h^2}\right)\left[-\left(\frac{F}{H_A A}\right)\left(\frac{2h}{\lambda_n^2}\right)\right] \cos\left(\lambda_n\frac{x_3}{h}\right) \mathrm{e}^{-\lambda_n^2\frac{H_A k}{h^2}t} \\
&= \sum_{n=1}^{\infty}\left(\frac{2F}{Ah}\right) \cos\left(\lambda_n\frac{x_3}{h}\right) \mathrm{e}^{-\lambda_n^2\frac{H_A k}{h^2}t}
\end{aligned}
\qquad (9.44)
$$

Integrating Eq. (9.44) term by term yields

$$
p(x_3,t) = \left[\sum_{n=1}^{\infty}\left(\frac{2F}{Ah}\right)\left(\frac{h}{\lambda_n}\right) \sin\left(\lambda_n\frac{x_3}{h}\right) \mathrm{e}^{-\lambda_n^2\frac{H_A k}{h^2}t}\right] + f(t) \qquad (9.45)
$$

where $f(t)$, a "constant" of integration with respect to x_3, is determined from the boundary condition

$$p(0,t) = 0 \tag{9.46}$$

Applying Eq. (9.46) to Eq. (9.45), we see that $f(t) = 0$. Hence,

$$p(x_3,t) = \sum_{n=1}^{\infty} \left(\frac{2F}{\lambda_n A}\right) \sin\left(\lambda_n \frac{x_3}{h}\right) e^{-\lambda_n^2 \frac{H_A k}{h^2} t} \tag{9.47}$$

Let us briefly discuss the interpretation of Eqs. (9.40) and (9.47). At $t = 0^+$, $u_3(0,t) = 0$; there is no instantaneous deformation. This reflects the fact that fluid cannot be instantaneously squeezed from the tissue. The external load is initially supported entirely by a hydrostatic pressure equal to $\phi^f p$ in the fluid and $\phi^s p$ in the solid, except at $x_3 = 0$ where the boundary condition requires $p = 0$ (see Figure 9.2). Furthermore, the pressure gradient, $\partial p/\partial x_3$, during fluid flow is greatest at the top surface and declines with depth in the tissue. Accordingly, v^f and u_3 are initially greatest at the surface region (~top 10%, i.e., $x_3/h = 0.1$) where the solid has a larger load bearing role. With increasing time, the pressure gradient and fluid velocity become more uniform throughout the tissue. Compaction of the solid matrix, and increasing solid stress, proceed downward. Concomitantly, the fluid pressure declines.

The several curves in the graph on the right of Figure 9.2 correspond to different time points: $t = 0$ (solid line), $t = 0.001t_D$ (*), $t = 0.01t_D$ (solid line), $t = 0.1t_D$ (+), $t = 0.2t_D$ (–), $t = 0.4t_D$ (x), $t = 0.8t_D$ (Δ), $t = t_D$ (–), $t = 5t_D$ (○).

Figure 9.3 shows how the load is shared between the fluid and solid phases during confined compression. Notice that the total normalized stress is always one, i.e., the total stress is the sum of the stress in the solid phase and the stress in the fluid phase. As shown in Figures 9.2 and 9.3, fluid pressure declines over time to zero, such that at equilibrium the entire stress is borne by the solid matrix. Also note that load is shared differently depending on vertical location within the tissue.

9.7.2 Stress Relaxation

For stress relaxation, an initial displacement is applied and held constant (see Figure 7.2). Accordingly, the boundary condition at the tissue surface is

$$u_3(0,t) = -u^0 \tag{9.48}$$

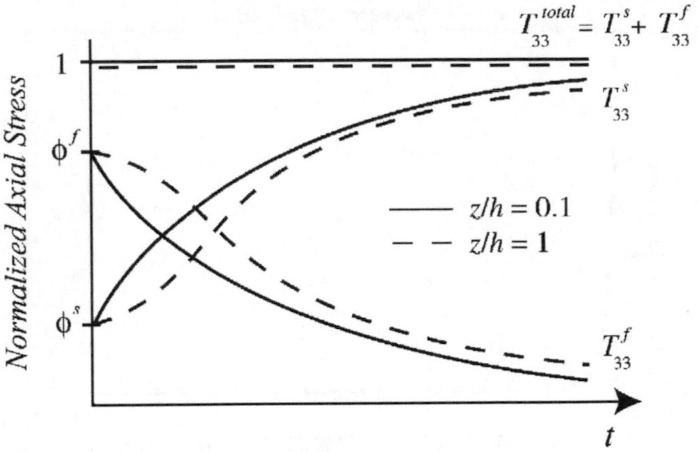

FIGURE 9.3: Load sharing between fluid and solid during a confined compression creep experiment.

The boundary condition at the bottom of the confinement chamber and the initial condition are still given by Eqs. (9.34) and (9.33), respectively. The solution to Eq. (9.32) under these conditions is

$$u_3(x_3,t) = \frac{u^0}{h}(x_3 - h) + 2u^0 \sum_{n=1}^{\infty} \frac{1}{\lambda_n} \sin\left(\lambda_n \frac{x_3}{h}\right) \exp\left(-\lambda_n^2 \frac{H_A k}{h^2} t\right) \tag{9.49}$$

Note: Here, $\lambda_n = n\pi$.

Similar to displacement at the surface in a creep test, the stress on the tissue's surface is experimentally accessible through the use of a standard materials testing system. Figure 9.4 shows the stress at the tissue's surface, which is given by Eq. (9.36) as,

$$\begin{aligned} T_{33}^s(0,t) &= H_A \left.\frac{\partial u_3}{\partial x_3}\right|_{(0,t)} = H_A \frac{u^0}{h} + H_A \frac{2u^0}{h} \sum_{n=1}^{\infty} e^{-\lambda_n^2 \frac{H_A k}{h^2} t} \\ &= H_A \frac{u^0}{h}\left[1 + 2\sum_{n=1}^{\infty} e^{-\lambda_n^2 \frac{H_A k}{h^2} t}\right] \end{aligned} \tag{9.50}$$

Note that the infinite stress at $t = 0$ is not a physical reality.

The equilibrium stress at the tissue surface is also of interest, given by

$$T_{33}^s(0,\infty) = H_A \frac{u^0}{h} = H_A \varepsilon \tag{9.51}$$

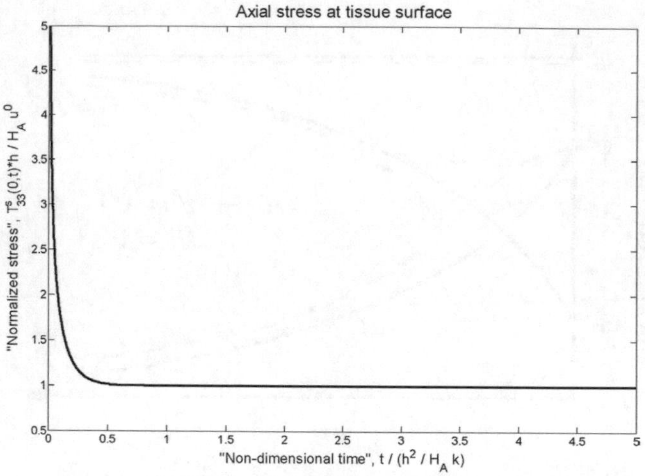

FIGURE 9.4: Stress at tissue's surface during a confined compression stress relaxation experiment.

Here, we use ε to denote the applied strain. At equilibrium, the result is similar to Hooke's law, with H_A as the "equivalent modulus of elasticity." Similar to creep, H_A is determined by the equilibrium response, and k is determined by curve fitting the experimental data. Note that the time constant governing stress relaxation is the same as that governing creep, i.e., the "gel diffusion time". Hence, when performing a creep or stress-relaxation test, equal amounts of time should be given.

The above solution applies only to an instantaneously applied displacement. In reality, it is experimentally impossible to apply a step displacement. Typically, a ramp displacement is applied as follows

$$u_3(0,t) = \begin{cases} v_0 t & 0 \leq t \leq t_0 \\ v_0 t_0 & t \geq t_0 \end{cases} \tag{9.52}$$

So, the stress-relaxation test is divided into a compressive phase and relaxation phase. The solution for stress-relaxation subject to ramp loading is beyond the scope of this text, and we refer the interested reader to reference [31] for the solution.

9.8 UNCONFINED COMPRESSION

Having analyzed the confined compression problem, let us now turn to another common method for testing biological tissues in compression, namely, unconfined compression. The unconfined compression setup is shown in Figure 9.1. Unconfined compression is most easily analyzed using cylindrical coordinates, r, θ, and z. Furthermore, the geometry of the testing setup suggests axisymmetric displacement and pressure fields, which means that the problem is independent of θ.

Assuming that the tissue specimen is compressed between two impermeable, frictionless platens, 1) the radial displacement and pressure are independent of z, and 2) the axial strain is independent of z [32]. Mathematically, these assumptions are

$$u_r = u_r(r,t), \quad p = p(r,t) \quad \text{and} \quad u_z = z\,E_{zz}(t) \tag{9.53}$$

From the continuity equation and the fact that, due to symmetry, $v_r^f(r = 0, t) = v_r^s(r = 0, t) = 0$, we get the following expression for the velocity of the fluid in the radial direction (see problem 8)

$$v_r^f = -\frac{r}{2}(1+\alpha)\frac{\partial E_{zz}}{\partial t} - \alpha v_r^s \tag{9.54}$$

where $\alpha = \dfrac{\phi^s}{\phi^f}$. Furthermore, as done before, pressure can be eliminated from the equations for the stress in the radial direction of the fluid and solid. Finally, combining that result with Eq. (9.54), one obtains the governing equation for radial displacement of the solid matrix during unconfined compression (see problem 8)

$$\frac{\partial^2 u_r}{\partial r^2} + \frac{1}{r}\frac{\partial u_r}{\partial r} - \frac{u_r}{r^2} - \frac{1}{H_A k}\frac{\partial u_r}{\partial t} = \frac{1}{H_A k}\frac{r}{2}\frac{\partial E_{zz}}{\partial t} \tag{9.55}$$

Note that this is a nonhomogeneous PDE in cylindrical coordinates. Though complex, this equation can now be solved using Laplace transforms under conditions of creep and stress relaxation provided appropriate boundary and initial conditions and the axial strain rate. For boundary conditions we have the following, where the radius of the specimen is a,

$$\begin{aligned} T_{rr}^{\text{total}}(r = a, t) &= 0 \\ p(r = a, t) &= 0 \end{aligned} \tag{9.56}$$

both of which must be transformed to displacement boundary conditions. For stress relaxation the axial strain rate is prescribed as either a step (i.e., Heaviside) or a ramp displacement, where for creep we have

$$\int_0^{a(t)} T_{zz}^{\text{total}} 2\pi r \mathrm{d}r = F(t) \tag{9.57}$$

a Heaviside load application, i.e., $F(t) = -F_0H(t)$, determining the axial strain rate. For infinitesimal deformations, the radius does not change over time, as the initial and final configurations of the tissue are approximately the same. The initial condition is

$$u_r(r,0) = 0 \tag{9.58}$$

Solutions for each of these situations were first worked out by Armstrong and colleagues in 1984 [32].

Lastly, we would be remiss to not mention creep indentation. For the interested reader, more information on creep indentation can be found in the references [33,34].

9.9 PROBLEMS

1. A biphasic material has a porosity of 96%. What is the solidity? If the true density of the fluid is 1 g/mL, what is the apparent density of the fluid?
2. For the biphasic theory, explain how the solid portion of the matrix is compressible while the biphasic material is incompressible.
3. The constitutive relationship for the solid portion of a biphasic material is $T_{ij}^s = -\phi^s p\delta_{ij} + \lambda E_{kk}\delta_{ij} + 2\mu E_{ij}$. Describe in words what each term on the right hand side represents.
4. This problem will examine confined compression.
   (a) Using the continuity equation, Eq. (9.10), show that $v_3^f = -\dfrac{\phi^s}{\phi^f}v_3^s = -\left(\dfrac{1-\phi^f}{\phi^f}\right)v_3^s$. (Hint: Consider the boundary conditions). What is $v_3^f - v_3^s$?
   (b) Derive the governing equation for confined compression, Eq. (9.32), following the steps below. Let the x_3 axis be coincident with the loading axis. There is no motion allowed in the x_1 or x_2 directions.
      (i) Using Eq. (9.31), write the solid and fluid stress tensors, $\underline{T}^s$ and $\underline{T}^f$. Let $H_A = \lambda + 2\mu$ (H_A is known as the aggregate modulus).
      (ii) Write the conservation of momentum equations in the x_3 direction for the fluid and solid phases under static conditions.
      (iii) Use the last result from part (a) and the fact that $\phi^s + \phi^f = 1$ to eliminate p between the conservation of momentum equations. The result is Eq. (9.32), where $k = \dfrac{(\phi^f)^2}{K}$.
5. A force sensor of small area A is placed on the side wall inside of the confined compression chamber at a distance midway between the tissue surface and the bottom of the chamber. What is the approximate force recorded by the sensor? How could the actual (not approx.) force be calculated? Explain why this force develops.

6. Consider the following data from a confined compression creep experiment of a tissue-engineered articular cartilage construct with thickness $h = 0.5$ mm. A step load of $F = 0.007$ N was applied to a porous platen of $A = 0.503$ mm^2. Find H_A and k. The latter requires nonlinear curve fitting. Matlab©'s "cftool" command is one possible method. Use Eq. (9.40) to fit the curve using $n = 1$, 5, and 10 terms of the sum. You may find that the data and fit do not match. If so, why may this be?

TIME(s)	DISPLACEMENT (mm)
0	0
1	0.0052
2	0.0071
3	0.0082
4	0.009
7	0.0107
10	0.0116
15	0.0128
20	0.0135
25	0.0142
40	0.015
60	0.016
80	0.017
120	0.018
140	0.0183
240	0.0192
340	0.0199
390	0.02
450	0.0201
470	0.0201
500 ($\approx \infty$)	0.0201

7. Verify that Eq. (9.49) satisfies the boundary and initial conditions of stress relaxation, Eqs. (9.34), (9.48), and (9.33). Also, evaluate Eq. (9.49) at $t = \infty$ to obtain the equilibrium response.

The formula for the coefficients of a Fourier sine series, $c_n = \frac{2}{h}\int_0^{\infty} f(x)\sin\left(\frac{n\pi x}{h}\right)dx$, where $f(x) = \sum_{n=1}^{\infty} c_n \sin\left(\frac{n\pi x}{h}\right)$, will be helpful.

8. This problem will examine unconfined compression.
   (a) Use the continuity equation, Eq. (9.9), in cylindrical coordinates, appropriate boundary conditions, and the fact that $v_z^f = v_z^s$ (because there is no relative motion between the fluid and solid in the z direction) to arrive at Eq. (9.54). Recall that unconfined compression is an axissymmetric situation.

   In cylindrical coordinates, $\underset{\sim}{E}$ has the following matrix representation

$$[\underset{\sim}{E}] = \begin{bmatrix} \frac{\partial u_r}{\partial r} & \frac{1}{2}\left(\frac{1}{r}\frac{\partial u_r}{\partial \theta} + \frac{\partial u_\theta}{\partial r} - \frac{u_\theta}{r}\right) & \frac{1}{2}\left(\frac{\partial u_r}{\partial z} + \frac{\partial u_z}{\partial r}\right) \\ \frac{1}{2}\left(\frac{1}{r}\frac{\partial u_r}{\partial \theta} + \frac{\partial u_\theta}{\partial r} - \frac{u_\theta}{r}\right) & \frac{1}{r}\frac{\partial u_\theta}{\partial \theta} + \frac{u_r}{r} & \frac{1}{2}\left(\frac{\partial u_\theta}{\partial z} + \frac{1}{r}\frac{\partial u_z}{\partial \theta}\right) \\ \frac{1}{2}\left(\frac{\partial u_r}{\partial z} + \frac{\partial u_z}{\partial r}\right) & \frac{1}{2}\left(\frac{\partial u_\theta}{\partial z} + \frac{1}{r}\frac{\partial u_z}{\partial \theta}\right) & \frac{\partial u_z}{\partial z} \end{bmatrix}$$

   Furthermore, recall Eq. (6.41) and (6.42).

   (b) Write the radial equilibrium equations for the fluid and the solid constituents in terms of displacements.
   (c) Now, eliminate the pressure term from the two equations and use the result from part (a) to arrive at Eq. (9.55).

• • • •

Acknowledgments

We would like to extend wholehearted gratitude to the late Professor R. Skalak and to Professor W. Michael Lai, both of Columbia University, whose biomechanics classes were taken by Professor Athanasiou during his time as a graduate student. We would also like to acknowledge the book *Introduction to Continuum Mechanics* coauthored in part by Professor Lai. Much of the first half of this text is modeled off of that excellent treatise, which taught Professor Athanasiou continuum mechanics many years ago.

We would also like to thank prior teaching assistants of the Introduction to Continuum Biomechanics course that has been taught at Rice University for some time now. Specifically, Drs. Michael Detamore, Adrian Shieh, and C. Corey Scott were instrumental in editing and adding to the course notes. Graduate student Najmuddin Gunja helped create the first soft copy of the course notes, which is reflected in the present text. Finally, thanks to the many students who have sweated through the course.

Afterword

We hope you have learned much about the subject of continuum biomechanics from reading this book. However, reading is not enough! There is no substitute for solving continuum mechanics problems. In fact, much of the enrichment of your study of biomechanics is contained in the problems we have included at the end of each chapter. As this is an introductory text, we have presented a breadth of topics, but depth on only a few. However, as with all subjects, there is much more to learn. The subject of continuum biomechanics could fill up volumes, even at an elementary level. We have chosen to curtail our discussion to salient features of the theories we have addressed, presentation and explanation of the constitutive equations, and demonstration of important points by solving some simple, classical problems.

We began the text with a solid foundation in Cartesian tensors. Note that curvilinear coordinates and a presentation of general tensor algebra and calculus are the subjects of intermediate and advanced texts. We then spent a good deal of effort studying strain (kinematics) and stress separate from one another, as they are distinct concepts. Finally, we showed how stress and strain can be related to each other through the many constitutive theories discussed.

Along the way, we solved several problems. If you take a moment to reflect, there is a relatively straightforward recipe for tackling continuum mechanics problems. One begins with conservation laws which must be true for all systems (e.g., conservation of mass, linear and angular momentum, and energy). One then substitutes the constitutive equation(s) describing the body of interest and the kinematic relations (i.e., compatibility equations) into the conservation laws to arrive at governing equations for the loading configuration under investigation. Finally, appropriate boundary and initial conditions are applied to obtain a solution to the governing equations, typically yielding displacement or velocity fields. From these fields, strains and strain rates can be calculated. Lastly, stress is determined from substituting the strains or strain rates back into the constitutive equations. Reaction forces can be obtained by integrating the stress over its area of application.

Though the physical concepts and solution approach are straightforward, the mathematical aspects of the solution of continuum biomechanics problems can be quite complex. While closed-form solutions are desirable, numerical approaches are often taken (and are sometimes the only recourse, as not every problem has a closed form solution). A common numerical approach in continuum mechanics is the finite element method. A fantastic introductory text on finite element analysis is reference [35]. In fact, the references contain many texts that expand upon the material presented here at an intermediate or advanced level, including the mathematics necessary to solve more complicated problems. We have made every effort to cite texts that one should be able to pick up and follow after reading this book (including the fact that many of the references share our notation).

Finally, we have attempted to integrate the vertical and horizontal layers of complexity among the theories we have presented. For example, we view hyperelasticity as vertically above linear elasticity, as the latter is a subset of the former. Similarly, we showed how linear viscoelasticity is vertically above linear elasticity and Newtonian viscous fluids. While elasticity is a subset of poroelasticity or thermoelasticity, we view these as more horizontally related to elasticity through the inclusion of different aspects of physical reality (i.e., not solely stress or strain, but fluid and thermal effects as well). Mixture theory is somewhat in a class of its own, though "mixtures of one component" certainly reduce to the theories of fluids or solids. However, we showed how biphasic theory and poroelasticity are related. We would like to point out, that while these theories provide great mathematical models of physical reality, in that they are able to accurately describe a vast array of material behavior, they are only *models* of physical reality, and are hence limited by the extent to which they sufficiently describe certain aspects of material behavior.

Though the subject of this book is biomechanics, our emphasis was the foundations of continuum mechanics. We presented theories that are germane to establishing descriptions of and solutions for the behavior of biological tissues and materials, without being exhaustive in the "bio" aspects. We believe that once these principles and theories have been studied and understood, it is with relative ease that one can transition to biomechanics problems.

While we have made every effort to ensure the content of the text, homework problems, and homework solutions are correct, "to err is human." We would appreciate receiving any and all constructive feedback as to how we can improve the text, as well as bringing to our attention any errors of commission or omission that you may come across.

Thank you.

KA2 and RMN

Bibliography

[1] Taber LA. Nonlinear Theory of Elasticity Applications to Biomechanics. River Edge, World Scientific Publishing Co. Pte. Ltd. 2004.

[2] Lai MW, Rubin D, Krempl E. Introduction to Continuum Mechanics. 3rd ed. Burlington, Butterworth Heinmann 1993.

[3] Spencer AJM. Continuum Mechanics. Mineola, Dover Publications, Inc. 2004.

[4] Basar Y, Weichert D. Nonlinear Continuum Mechanics of Solids—Fundamental Mathematical and Physical Concepts. New York, Springer Verlag 2000.

[5] Holmes MH, Mow VC. The nonlinear characteristics of soft gels and hydrated connective tissues in ultrafiltration. J Biomech 1990;23:1145–56. doi:10.1016/0021-9290(90)90007-P

[6] Veronda DR, Westmann RA. Mechanical characterization of skin-finite deformations. J Biomech 1970;3:111–24. doi:10.1016/0021-9290(70)90055-2

[7] Pioletti DP, Rakotomanana LR, Benvenuti JF, Leyvraz PF. Viscoelastic constitutive law in large deformations: Application to human knee ligaments and tendons. J Biomech 1998;31:753–7. doi:10.1016/S0021-9290(98)00077-3

[8] Miller K. Method of testing very soft biological tissues in compression. J Biomech 2005; 153–58.

[9] Maurel W, Wu Y, Thalmann NM, Thalmann D. Biomechanical Models for Soft Tissue Simulation. New York, Springer 1998.

[10] Leipholz H. Theory of Elasticity. Leyden, Noordhoff International Publishing 1974.

[11] Truskey GA, Yuan F, Katz DF. Transport Phenomena in Biological Systems. Upper Saddle River, Pearson Prentice Hall 2004.

[12] Haberman R. Applied Partial Differential Equations with Fourier Series and Boundary Value Problems. 4th ed. Upper Saddle River, Pearson Prentice Hall 2004.

[13] Reddy JN. An Introduction to Continuum Mechanics with Applications. New York, Cambridge University Press 2008.

[14] Fung YC. Biomechanics Mechanical Properties of Living Tissues. 2nd ed. New York, Springer Verlag 1993.

[15] Mazumdar JN. Biofluid Mechanics. River Edge, World Scientific Publishing Co. Pte. Ltd. 1992.

[16] Humphrey JD, Delange SL. An Introduction to Biomechanics: Solids and Fluids, Analysis and Design. New York, Springer Science+Business Media, Inc. 2004.

[17] Machiraju C, Phan AV, Pearsall AW, Madanagopal S. Viscoelastic studies of human subscapularis tendon: Relaxation test and a Wiechert model. Comput Methods Programs Biomed 2006;83:29–33. doi:10.1016/j.cmpb.2006.05.004

[18] Boehler JP, Betten J, Spencer AJM. Applications of Tensor Functions in Solid Mechanics. New York, Springer Verlag 1987.

[19] Flugge W. Viscoelasticity. Waltham, Blaisdell Publishing Co. 1967.

[20] Haddad YM. Viscoelasticity of Engineering Materials. New York, Chapman & Hall 1995.

[21] Lakes RS. Viscoelastic Solids. Boca Raton, CRC Press LLC 1999.

[22] Koay EJ, Shieh AC, Athanasiou KA. Creep indentation of single cells. J Biomech Eng 2003;125:334–41. doi:10.1115/1.1572517

[23] Biot MA. General theory of three-dimensional consolidation. J Appl Phys 1941;12:155–64. doi:10.1063/1.1712886

[24] Wang HF. Theory of Linear Poroelasticity with Applications to Geomechanics and Hydrogeology. Princeton, Princeton University Press 2000.

[25] Simon BR. Multiphase poroelastic finite element models for soft tissue structures. Appl Mech Rev 1992;45:191–218.

[26] Norris A. On the correspondence between poroelasticity and thermoelasticity. J Appl Phys 1992;71:1138–41. doi:10.1063/1.351278

[27] Nowacki W. Thermoelasticity. 2nd ed. Elmsford, Pergamon Press Inc. 1986.

[28] Parkus H. Thermoelasticity. Waltham, Blaisdell Publishing 1968.

[29] Atkin RJ, Craine RE. Continuum theories of mixtures: Basic theory and historical development. Q JMAM 1976;29:209–44. doi:10.1093/qjmam/29.2.209

[30] Humphrey JD. Continuum biomechanics of soft biological tissues. Proc R Soc Lond A 2003;459:3–46. doi:10.1098/rspa.2002.1060

[31] Mow VC, Kuei SC, Lai WM, Armstrong CG. Biphasic creep and stress relaxation of articular cartilage in compression: Theory and experiments. J Biomech Eng 1980;102:73–84.

[32] Armstrong CG, Lai WM, Mow VC. An analysis of the unconfined compression of articular cartilage. J Biomech Eng 1984;106:165–73.

[33] Mak AF, Lai WM, Mow VC. Biphasic indentation of articular cartilage—I. Theoretical analysis. J Biomech 1987;20:703–14. doi:10.1016/0021-9290(87)90036-4

[34] Mow VC, Gibbs MC, Lai WM, Zhu WB, Athanasiou KA. Biphasic indentation of articular cartilage—II. A numerical algorithm and an experimental study. J Biomech 1989;22: 853–61. doi:10.1016/0021-9290(89)90069-9

[35] Ottosen NS, Petersson H. Introduction to the Finite Element Method. Harlow, Pearson Education Limited 1992.

Author Biography

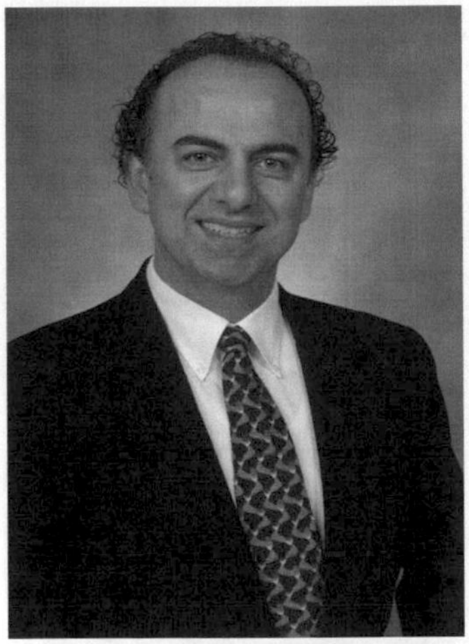

K.A. Athanasiou is the Karl F. Hasselmann professor of bioengineering at Rice University and an adjunct professor of orthopaedics and oral and maxillofacial surgery at the University of Texas. He also heads the Musculoskeletal Bioengineering Laboratory at Rice University. He holds a Ph.D. in mechanical engineering (bioengineering) from Columbia University.

R.M. Natoli received his Ph.D. degree from Professor Athanasiou's laboratory. He is currently on clinical rotations at Baylor College of Medicine as part of the MSTP program. His research project focused on impact loading of articular cartilage and mechanobiological aspects of articular cartilage tissue engineering. He also holds a B.S. in biological chemistry and A.B. in chemistry from the University of Chicago.